Western Michigan University

Chemistry I Laboratory Manual

3rd Edition

Sherine O. Obare | Brianna N. Hyder
Megan L. Grunert | Ekkehard Sinn

Australia • Brazil • Japan • Korea • Mexico • Singapore • Spain • United Kingdom • United States

Western Michigan University: General Chemistry I Laboratory Manual, 3rd Edition

Senior Manager, Student Engagement:
Linda deStefano
Janey Moeller

Manager, Student Engagement:
Julie Dierig

Marketing Manager:
Rachael Kloos

Manager, Production Editorial:
Kim Fry

Manager, Intellectual Property Project Manager:
Brian Methe

Senior Manager, Production and Manufacturing:
Donna M. Brown

Manager, Production:
Terri Daley

ALL RIGHTS RESERVED. No part of this work covered by the copyright herein may be reproduced, transmitted, stored or used in any form or by any means graphic, electronic, or mechanical, including but not limited to photocopying, recording, scanning, digitizing, taping, Web distribution, information networks, or information storage and retrieval systems, except as permitted under Section 107 or 108 of the 1976 United States Copyright Act, without the prior written permission of the publisher.

For product information and technology assistance, contact us at
Cengage Learning Customer & Sales Support, 1-800-354-9706
For permission to use material from this text or product, submit all requests online at **cengage.com/permissions**
Further permissions questions can be emailed to
permissionrequest@cengage.com

This book contains select works from existing Cengage Learning resources and was produced by Cengage Learning Custom Solutions for collegiate use. As such, those adopting and/or contributing to this work are responsible for editorial content accuracy, continuity and completeness.

Compilation © 2017 Cengage Learning
ISBN-13: 9781337782081

WCN: 01-100-101

Cengage Learning

5191 Natorp Boulevard
Mason, Ohio 45040
USA

Cengage Learning is a leading provider of customized learning solutions with office locations around the globe, including Singapore, the United Kingdom, Australia, Mexico, Brazil, and Japan. Locate your local office at:
international.cengage.com/region.

Cengage Learning products are represented in Canada by Nelson Education, Ltd.
For your lifelong learning solutions, visit **www.cengage.com/custom.**
Visit our corporate website at **www.cengage.com.**
Visit signature labs online at **signaturelabs.com.**

Printed at CLDPC, USA, 05-20

Contents

Laboratory Techniques: Safety Precautions 1

1. Size and Scale 25
2. Line Spectra: Evidence for Atomic Structure 31
3. Molecular Representations 39
4. Identification of a Compound by Mass Relationships 55
5. Determination of a Chemical Formula 63
6. Solubility 71
7. Types of Chemical Reactions 83
8. Limestone Caves 93
9. Analysis of a Carbonated Beverage 107
10. The Fuel in a Bic® Lighter 121

TECH 0430

Laboratory Techniques: Safety Precautions

Prepared by Norman E. Griswold, Nebraska Wesleyan University

As every chemist knows, there are many potential hazards in chemistry laboratories, some of which can be serious. However, chemists can avoid accidents by being thoroughly familiar with appropriate experimental techniques, by knowing the hazards associated with the chemicals they use, and by taking common sense safety precautions. Unlike most research and production laboratories, a school laboratory contains a relatively large number of students working, many without extensive knowledge of laboratory techniques, equipment, chemical hazards, or safety precautions. Consequently, students usually need some initial instruction in order to help them develop the necessary knowledge and awareness to avoid accidents.

By studying this module, you will learn basic safety rules and procedures applicable to chemistry laboratories. There is also a section describing what you should do if an accident occurs. On the last page, there is a Laboratory Safety Agreement that you must sign before you perform any experiments in the laboratory.

I. GENERAL SAFETY RULES

A. For Personal Protection

1. Wear eye protection at all times in the laboratory.

    All staff, students, and visitors in the laboratory must wear proper eye-protection glasses or goggles that are approved by the appropriate authorities. Many states have eye-protection laws and require use of splashproof goggles. Your eye protection should protect both the front and sides of your eyes, should meet Occupational Safety and Health Administration (OSHA) standards, and should fit over your prescription glasses, if you wear them.

    Normally, you should not wear contact lenses in the laboratory. If they are permitted and you do wear them, you must wear *fitted* goggles at all times. Contact lenses are hazardous, because they concentrate vapors under the lens, trap foreign matter, and interfere with the effectiveness of eyewash fountains. Soft contact lenses are particularly hazardous, because they can absorb and retain chemical vapors. Ordinary eyeglasses or sunglasses are not adequate protection in the laboratory.

© 1993 Cengage Learning. ALL RIGHTS RESERVED. No part of this work covered by the copyright herein may be reproduced, transmitted, stored, or used in any form or by any means graphic, electronic, or mechanical, including but not limited to photocopying, recording, scanning, digitizing, taping, Web distribution, information networks, or information storage and retrieval systems, except as permitted under Section 107 or 108 of the 1976 United States Copyright Act, without the prior written permission of the publisher.

2. Wear sensible clothing and tie back long hair.

In the chemistry laboratory, "sensible" clothing means old clothes that are fire resistant and not too loose, especially the sleeves. You should remove neckties, scarves, and jewelry. It is a good idea to wear a laboratory coat or apron.

"Sensible" clothing also includes shoes that cover your entire foot. Sandals and high heels are not safe. Shoes provide initial foot protection from dropped containers and spilled chemicals.

Keep all extra clothing, such as coats, hats, and scarves, off your laboratory bench and, if possible, out of the laboratory altogether.

If you have long hair, it can easily catch fire when you are close to an open flame. Usually this happens when you bend forward and your hair falls in front of your shoulders. Hair sprays only make matters worse. Consequently, you should fasten your long hair back.

3. Avoid absorbing, inhaling, or ingesting chemicals while you are in the laboratory.

There are three main routes for entry of chemicals into the body.

(1) The respiratory tract (lungs): Vapors, mists, smoke, and even dust particles can carry lethal amounts of many substances into your body through your lungs. You can detect the presence of some toxic vapors by their noticeable odors, but many have no odor at all. For example, mercury metal has no odor, but its vapor pressure at room temperature allows mercury to enter the atmosphere at about one hundred times its threshold toxicity limit. To prevent inhaling toxic fumes, use the fume hood as your laboratory instructor directs.

If you are directed to detect odorous fumes by smell, do so with great care. Be certain that the reaction has ceased before you remove the container holding the reaction mixture from the fume hood. Place your hand near the container and gently fan some of the vapor toward you. Sniff carefully; do not inhale deeply. Return the container to the fume hood.

(2) The digestive tract: Never put anything in your mouth while you are in the laboratory. This precaution applies not only to chemicals, your fingers, and pipets, but also to beverages, food, and cigarettes. Do not bring food or drink into the laboratory, because it might become contaminated. Never put glassware in your mouth. Never use mouth suction to fill a pipet. Instead, your laboratory instructor will demonstrate the proper procedure for filling a pipet, using a rubber bulb.

(3) The skin: Some toxic substances can enter your body through your skin. Such substances include phenol, nitrobenzene, carbon tetrachloride, and solutions of cyanides. Some liquids, such as acids and bromine, can burn or otherwise damage your skin. Because many of these potentially hazardous liquids look like water, it is easy to forget their toxic or corrosive properties. Your cleanliness habits in the laboratory are an important part of skin-related accident prevention. Clean up spills immediately. Thoroughly wash your skin or any clothing that contacts a chemical. Periodically wash your hands and arms while working in the laboratory. Wash your hands and face with soap or detergent at the end of the laboratory period.

B. Working in the Laboratory

1. Do not attempt any unauthorized or altered experiments.

    Consult your laboratory instructor before you make any change in an experimental procedure. Based on experience and knowledge, your instructor may be able to foresee possible hazardous results due to the altered procedure.

2. Know where safety equipment is located and how to use it.

    Safety equipment includes fire extinguishers, safety showers, eyewash fountains, fire blankets, and an alarm or emergency telephone system. Your laboratory instructor will identify and describe the use of all these items during your first laboratory period. In addition, your instructor will clearly indicate the locations of all laboratory exits and describe an emergency evacuation plan. If such an introduction is not presented during your first laboratory period, ask your laboratory instructor to do so.

3. Never work alone in the laboratory.

    Your laboratory instructor must be within your sight and hearing range at all times while you are working in the laboratory. If you encounter difficulties, your instructor must be able to assist you.

4. Use the fume hood when necessary.

    Always perform experiments utilizing either toxic substances or substances with strong, irritating odors in a fume hood with proper airflow. Be certain that the hood is turned on. Do not place any portion of your body except your hands and lower arms into the hood.

    Among the substances that you should use exclusively in a fume hood are chlorine, bromine, formaldehyde, phenol, sulfur dioxide, hydrogen sulfide, carbon disulfide, glacial acetic acid, and concentrated aqueous ammonia. If you are using highly toxic or carcinogenic substances such as cyanides, benzene, or chloroform, use a fume hood with an airflow of 60–100 linear feet per minute.

    There are many other toxic substances. Therefore, it is imperative that you know the hazardous properties of *all* the substances you will use in an experiment. A major source of such information is the **Material Safety Data Sheet (MSDS)**, which must be available in the laboratory for every hazardous chemical you use. The MSDS sections on reactivity and health hazard data are of special interest for this purpose.

5. Dispose of waste and excess materials as directed by your laboratory instructor.

    Each person is responsible for handling waste and excess materials in ways that will minimize environmental contamination and personal hazard. You should be given clear procedures to follow for disposing of waste materials. Some general guidelines are:

    (1) Dispose of chemicals as directed by your laboratory instructor. Most reaction by-products and surplus chemicals must be neutralized or deactivated before disposal.

    (2) Promptly dispose of waste materials into the proper labeled container.

    (3) Place broken glass and porcelainware in separate containers specifically designated for this purpose. Ask for assistance before

disposing of a broken thermometer, because special handling is often required due to the mercury present in many thermometers.

(4) Place used matches and paper in a trash container, not in the sink. Dampen matches before disposing of them. Ask your laboratory instructor for disposal directions for any paper contaminated with a chemical.

II. GENERAL RULES FOR HANDLING EQUIPMENT

1. Only use equipment that is in good condition.

Defective equipment is a major accident source. Avoid using:

(1) beakers, flasks, funnels, graduated cylinders, or test tubes with chipped or broken rims;

(2) cracked beakers, flasks, graduated cylinders, test tubes, or crucibles;

(3) test tubes, beakers, or flasks with starshaped cracks at or near the bottom;

(4) beakers, flasks, or test tubes with severely scratched bases;

(5) burets, pipets, or funnels with chipped tips;

(6) glass tubing or glass rods with sharp edges;

(7) hardened rubber stoppers;

(8) screw clamps, buret clamps, or support rings with nonfunctioning parts.

2. Assemble your experimental apparatus carefully.

Keep your work space uncluttered by extra equipment and chemicals. Use only equipment that is dry, at least on the outside. Wet glassware is slippery and can easily be dropped and broken.

When you use a ring stand with a platform base, assemble the attached apparatus directly over the base, and not to one side. Be sure to keep the assembly back from the edge of the laboratory bench, and tighten the clamps firmly. When you use a Bunsen burner, be sure you can quickly move it away from the ring stand assembly. Place a wire gauze under any glassware that you are going to heat with a Bunsen burner.

3. Avoid touching hot objects.

Burns are among the most common accidents in chemistry laboratories. Use the following precautions to avoid burns.

Do not hold reaction vessels in your hand: many chemical reactions generate large amounts of heat. Place flat-bottomed reaction vessels on the laboratory bench before you mix chemicals in them. If you cannot place the vessel on a laboratory bench, attach a clamp to the vessel, and clamp it to a ring stand.

When you heat or mix chemicals in a test tube, do not hold the test tube in your hand. Hold it with a test-tube clamp, or place it in a beaker.

When you heat chemicals in a container, remember that you are applying heat to not only the chemical but also to the container and any clamp holding the container. Be very careful when you touch clamps or containers that have been heated or been near a heat source.

Heated glass seems to stay hot forever. Therefore, after you work with glass in or near a flame, lay the glass aside to cool. Use great caution when touching the glass after cooling. Protect your hands with a towel or heat-absorbing gloves if you must move hot equipment.

Do not lay hot glassware directly on the laboratory bench. The hot object may cause charring or other damage. Instead, place hot glass on a ceramiccentered wire gauze, altered as follows. Bend each corner of the gauze downward to form a 90° angle. Place the gauze on the bench so that the folded corners act as legs. This should keep the center of the gauze about 1 cm above the bench surface. Assume that anything on this gauze is hot. In this way, you can avoid many painful finger burns.

4. Use extreme caution when you insert glass into stoppers.

Occasionally, you must insert a piece of glass into a rubber stopper. If you force the glass into the stopper, the pressure can cause the glass to suddenly break and seriously injure your hand.

Use the following procedure when inserting glass tubing, glass rods, thermometers, funnels, thistle tubes, or other objects with small diameters into rubber stoppers.

Before beginning to insert glass tubing, check to be sure that the ends of the tubing are smooth. If the ends are rough, fire polish the tubing, and set it aside to cool before you try to insert it into a stopper. Next, check to see that the stopper is pliable when rolling it in your palm or squeezing it with your fingers. If your stopper is hard and inflexible, use a different one. Be certain that the stopper properly fits the intended opening. You will feel frustrated if you successfully insert a glass tube into a stopper, only to find that you used the wrong-size stopper. Also, make sure that the stopper hole is approximately the same diameter as the tubing you are inserting.

Using water or glycerine, lubricate the stopper hole and about 2 or 3 cm at the end of the tubing. Keep the rest of the tubing dry so that you can easily grip it with your hand. Wrap your hands with the opposite ends of a cloth. Hold the stopper in one wrapped hand. With the other wrapped hand, grasp the tubing near the lubricated end, as shown in Figure 1. Carefully insert the tubing into the stopper hole, using gentle pressure and a twisting motion. If the tubing does not slide through the hole when you exert slight pressure, apply additional lubricant to the tubing. *Never use force*. Usually, once you introduce the tubing into the hole, it will slide easily through the hole.

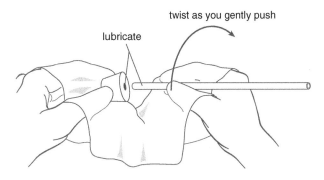

Figure 1
Inserting glass tubing into a stopper

Adjusting the tubing to the proper insertion depth is not difficult. Grasp the tubing as close to the stopper as possible. Continue to protect your hands with the cloth while you make the adjustment.

If you are inserting bent tubing, you may be inclined to grasp the bend in order to twist the tubing. You should avoid this, because the bend is generally the weakest part of a piece of bent tubing. The tubing will break if you put too much pressure on this point. Similarly, do not hold a thistle tube or a funnel by its top when inserting its stem into a rubber stopper.

III. HANDLING CHEMICALS

A. General Rules

1. Read labels carefully.

    (1) Be sure that you select the specified substances. Different chemicals can have similar names. Compare, for example, sodium nitrate and sodium nitrite, or stannic chloride and stannous chloride. The names of these compounds look quite similar, but their chemical behaviors are very different.

    (2) Use the specified concentration or form of all substances. If you use the wrong concentration of a solution, you may get unexpected or undesired results. Some solids are available as strips, wire, granules, or powder. In some experiments, it is vital that you use the correct solid reagent form.

    (3) Know the hazards associated with the substances you are using. To help warn users of hazards associated with specific chemicals, the National Fire Protection Agency (NFPA) has developed a diamond-shaped symbol for chemical labels that rates the hazard level for many chemicals. An example of the NFPA symbol is shown in Figure 2. The NFPA symbol has a red segment at the top that indicates flammability hazard. The left segment is blue and shows the toxicity hazard. The right segment is yellow and indicates reactivity. The bottom white segment is used to show special hazards, such as radioactivity, with a symbol of the hazard. If no specific hazard applies, the segment is left blank. Within each segment, there is a boldface black number that indicates the degree of hazard. The numerical ratings are:

    4 = extreme hazard (use goggles, gloves, protective clothing, and fume hood)
    3 = severe hazard
    2 = moderate hazard
    1 = slight hazard
    0 = no special hazard

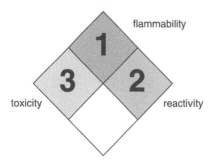

Figure 2
An example of the NFPA symbol for chemical hazards

2. Obtain only the specified amount of each substance.

Safety regulations strictly forbid your returning unused chemicals to their original containers. If you accidently returned a chemical to the wrong container, you might contaminate the entire contents of the container. Therefore, estimate the amount of each chemical you will need, and obtain only that amount from the reagent bottle. If it is difficult for you to estimate the proper amount, be conservative. You can always return for more if you need it.

If you obtain too much of a chemical, you must properly dispose of the excess. Consult your laboratory instructor for specific disposal instructions.

3. Leave chemicals in their proper places.

Bring your own container to the dispensing bottle on the reagent shelf. Do not take the dispensing bottle to your laboratory bench. It is not wise to carry large amounts of chemicals around the laboratory.

4. Clean up all spills immediately.

If you spill a chemical, it is your responsibility to immediately consult your laboratory instructor about the proper cleanup and disposal procedure. You are the only one who knows specifically what was spilled and where. The cleanup only takes a short time, and quick action can prevent further accidents.

If you spill a chemical on a balance, clean it up carefully and quickly. Otherwise, this delicate and expensive instrument may become corroded and permanently damaged.

Take care to remove any drops clinging to the outside of a reagent bottle after you have withdrawn reagent.

5. Label all containers to identify their contents.

When you obtain chemicals from the reagent shelf, be certain to label the containers into which you dispense the chemicals. Even if you plan to use the chemical immediately, you may have an excess that you must dispose of later. Don't trust your memory to keep track of such chemicals—use labels. You can use a pencil to write directly on the etched spot on many flasks and beakers. If you use pencil, you can easily erase these labels later.

B. Handling Liquids

1. Use proper technique when you obtain a liquid.

Take a labeled container to the reagent shelf. Transfer the liquid to your container at a nearby sink, in order to avoid spills on the bench, shelf, or floor.

From a dropper bottle: Be sure that you never touch your container or its contents with the dropper. Proper technique is shown in Figure 3. Never lay the dropper on any surface. Be careful to avoid contaminating the dropper bottle contents.

From a stoppered bottle: When you transfer the liquid, hold the stopper in your hand or place it upside down on a flat surface. Do not lay the stopper on its side. Proper technique is shown in Figure 4. If you follow this procedure, you will avoid contaminating the contents of the reagent bottle.

When you return the stopper to the reagent bottle, do not interchange stoppers. You may contaminate the reagent if you insert a stopper intended for another bottle.

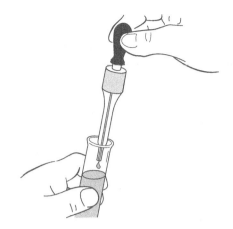

Figure 3
Transfer of a liquid using a dropper

When using a pipet to transfer liquid, pour some of the liquid into your own labeled container and pipet from that container into another. ***Do not pipet directly from a reagent bottle.***

From automatic dispensers: The apparatus has an automatic zero adjustment. You will dispense the liquid through a stopcock. Ask your laboratory instructor for specific directions for using this apparatus.

Another dispensing device you may use is the Repipet®. This device consists of a syringe pump mounted on top of a bottle. You can set the stop on the syringe to deliver a specified volume of liquid. To draw liquid into the pipet, lift the syringe plunger as far as the stop. Then, fully depress the plunger to deliver the specified amount of liquid from the delivery tube.

2. Be careful when you mix liquids.

When you mix liquid chemicals with water, always add the chemicals to the water, rather than vice versa. If you do this, the resulting solution will always be the more dilute and less hazardous solution. An accidental spattering from more dilute solution will be less

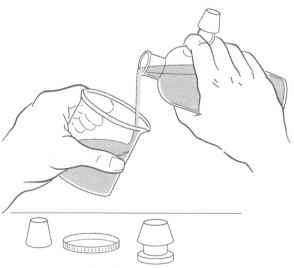

proper position of stoppers
left on bench

Figure 4
Pouring liquids from a bottle

8

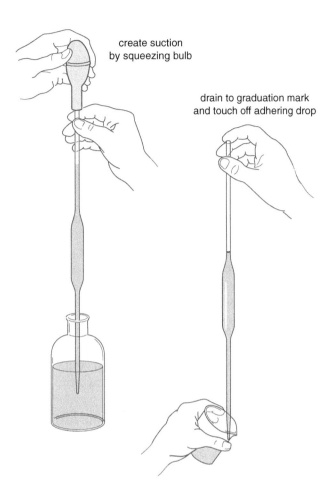

Figure 5
Filling a pipet

dangerous. Always add chemicals slowly, in small amounts. It is especially important to use this procedure when mixing an acid and water, so that the water can absorb the heat generated by the dilution process.

3. Use a rubber bulb when you pipet liquids.

   Never use your mouth to provide suction for filling a pipet. Instead, use a small rubber bulb at the top of the pipet to create suction to draw liquids into the pipet. The proper technique for filling a pipet is shown in Figure 5 and will be demonstrated by your laboratory instructor.

4. Heat liquids cautiously.

   Use a beaker or flask and a hot plate when heating a larger amount of liquid. Secure the container with a clamp.

   Heat small amounts of liquids in test tubes. Attach a test tube clamp to the upper part of the test tube. Place the lower part of the test tube in a beaker of boiling water.

   If you need to heat a liquid in a test tube to a temperature higher than 100 °C, you may heat the test tube and its contents using a hot plate and beaker containing a solvent, as directed by your laboratory instructor, and you must be extremely cautions. The liquid at the bottom of the test tube may boil more quickly than the rest, and the resulting expansion may cause hot liquid to spurt out the top of the test tube. Therefore, you should never look into a test tube or point the open end of the test tube toward anyone while you are heating a test tube in this manner.

Be extremely cautious whenever you heat *flammable* liquids. You should only heat flammable liquids in a steam bath or on a special hot plate; either way, you must be supervised by a laboratory instructor.

Never heat a liquid in a graduated cylinder or in any other volumetric glassware.

5. Dispose of liquids properly.

Dispose of liquids as specified by your laboratory instructor. Never pour flammable or water-immiscible liquids into the sink or other outlet that drains into the sewer. Transfer these liquids to appropriately labeled containers for disposal. Among the substances you must dispose of in this way are solvents such as acetone, ethyl alcohol, hexane, kerosene, methyl alcohol, and toluene.

C. Handling Solids

1. Transfer solids into widemouthed containers.

Solids are somewhat more difficult to transfer than are liquids. Take a labeled widemouthed container, such as a beaker, to the reagent shelf to obtain the solid you need. Make the transfer at or near the reagent shelf.

2. Transfer solids by rotating the reagent bottle.

During the transfer, hold the reagent bottle stopper in your hand or lay it on the bench, as shown in Figure 4, in order to prevent contamination.

Never insert your spatula directly into the reagent bottle. Instead, pour the chemical from the reagent bottle into your container. While pouring, slowly rotate the tipped bottle back and forth about an imaginary axis that passes through the top and bottom of the bottle,

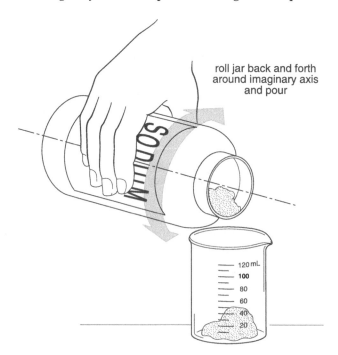

Figure 6
Rotating a reagent bottle in order to pour a solid safely

as shown in Figure 6. If you tip the bottle too far in an attempt to transfer the solid without rotating the bottle, large chunks may suddenly fall into your container, possibly causing a spill.

When you are finished, be sure to close the reagent bottle with the proper stopper. Do not interchange stoppers from other reagent bottles.

3. When you mix chemicals, add solids to liquids.

When mixing a solid with a liquid, add solid to liquid, rather than vice versa, with continuous stirring. Add solids in small amounts unless you are instructed otherwise.

4. Dispose of solids properly.

Dispose of solids as specified by your laboratory instructor. Place the solids in appropriately labeled containers for disposal.

IV. WHEN AN ACCIDENT OCCURS

Immediately attend to all physical and chemical injuries. Ask another student to report the accident to your laboratory instructor.

A. Burns

If you burn your skin with a hot object, flames, or chemicals, immediately flush the affected area with cold water. Continue flushing for 20 min.

If a large area of skin is burned and must be flushed, be careful not to get chilled.

Tell your laboratory instructor about the accident, even if the burn does not seem serious. Your instructor will decide whether or not you should seek medical attention.

In general, physicians prefer that you do not apply ointment to burns, especially serious ones. Often, they must remove the ointment because it can inhibit healing.

B. Splattered Chemicals

1. In your eyes.

You must wear eye protection at all times in the laboratory. Therefore, there should be no opportunity for chemicals to splatter in your eyes.

If a chemical splashes on your face while you are wearing splash-proof goggles or glasses, *do not remove your goggles*. Instead, drench your face and goggles at the nearest eyewash fountain. After you have removed all of the chemical from your goggles and face, take off your goggles, and drench the part of your skin where the goggles contact your face.

If a chemical does somehow get in your eye, immediately *YELL* for help in getting to an eyewash fountain. Drench your eye with water. Force your eye open, if necessary. Hold your eyelid away from your eyeball, and roll your eyeball so that the water flushes the entire eyeball. Continue to flush your eye with water for at least 15 min.

Ask your laboratory instructor whether or not you should seek medical attention.

2. On your skin.

For small spills on your skin, immediately rinse the affected area with large amounts of water. Remove any jewelry which might interfere with a thorough rinsing. Continue rinsing the area for 5–10 min.

If the spilled chemical is an acid, your laboratory instructor should follow the first water rinsing with a rinse of the sodium hydrogen carbonate solution kept in the laboratory for such an emergency. Resume rinsing with water for 5–10 min longer.

For spills of a base, your laboratory instructor should follow the first water rinsing with a rinse of boric acid solution, also found in the laboratory for emergency use. Resume rinsing with water for 5–10 min longer.

If a large area of your skin is affected, use the safety shower.

Ask your laboratory instructor whether or not you should seek medical attention.

3. On your clothing.

If possible, immediately remove all affected clothing. *Seconds count*, so don't hesitate because of modesty. Drench the skin contacted by the contaminated clothing with plenty of water. For large spills, immediately use the safety shower, regardless of any initial cold or discomfort. However, avoid becoming too chilled. Your laboratory instructor will supervise your use of the shower and tell you when to leave it.

Discard all clothing that is chemically contaminated, including shoes and belts, following the directions of your laboratory instructor. Change into clean, dry clothes as soon as possible.

Ask your laboratory instructor whether or not you should seek medical attention.

C. Fire

If you have long hair, you should tie it back while you are in the laboratory. If you do so, you are not likely to catch your hair on fire. However, if your hair does catch on fire, immediately *YELL* for help. Use your hands to keep the burning hair away from your face and drench the burning hair with water at the nearest sink. Your laboratory instructor and nearby students will help.

If your clothing is on fire, use the *STOP-DROP-ROLL-YELL* technique for putting out the flames. *STOP* what you are doing, *DROP* to the floor, and *ROLL* to try and smother the flames. *YELL* for others to help quickly extinguish the fire.

Whether it's your clothing or your hair on fire, **do not run** to a safety shower or a fire blanket. Running increases the burning rate and therefore the possibility that hot, toxic fumes will enter your lungs. Let someone else run and get the fire blanket. Also, safety showers are for chemical spills, not extinguishing fires!

If a fire blanket is used, your laboratory instructor will have it removed as soon as possible, in order to minimize skin burns from fused clothing. After the blanket has been removed, your laboratory instructor will supervise drenching you in the safety shower.

While you are receiving aid, another student should try to shut off or reduce the fuel supply to the original fire. If necessary, she/he should direct the spray from a fire extinguisher at the base of the fire. If the fire cannot be quickly extinguished, everyone should leave the laboratory. Then activate the building fire alarm, call the fire department, and notify the campus emergency services.

After the fire has been extinguished, tell your laboratory instructor which extinguishers were used, so that they can be tagged as empty and replaced with filled ones.

D. Injury

Thoroughly wash minor cuts, making certain to remove all foreign materials from the wound, such as traces of chemicals and broken pieces of glass. Apply a bandage to keep the wound clean and to avoid further irritation.

For major cuts and severe bleeding, fast action is critical. Have someone call for emergency medical aid. Have someone quickly apply direct pressure to the open wound with a clean compress or cloth, untill a tight bandage can be applied to control bleeding. When medical personnel arrive, they will decide what action to take. Only a medical professional is trained to treat serious injury.

_____   _____ _____
name   date   instructor
_____   _____
section   partner's name

Write your answers in pen and include units.

Laboratory Safety Quiz

1. Briefly describe the required protection and appropriate attire.

2. In your laboratory, where is the nearest:
   (1) fire extinguisher?

   (2) safety shower?

   (3) eyewash fountain?

   (4) location of the broken glass container?

   (5) telephone?

   (6) location of gloves (and the type)?

3. Briefly describe how you can avoid ingesting chemicals into your digestive tract.

4. What does MSDS mean?

5. Briefly explain what you must do before you dispose of waste or excess materials.

6. List the defects that make glassware unusable.

7. Briefly describe the precautions you should take when you mix chemicals in a test tube.

8. Briefly describe how you should protect your hands while you are inserting glass tubing into a stopper.

9. What does the upper segment of the NFPA symbol indicate?

10. Briefly describe the proper technique for heating liquids in test tubes to temperatures below 100 °C.

11. Briefly explain what you should do *first* if you burn a small area on your skin in the laboratory.

12. What should you do immediately if your clothing starts burning?

13. When diluting an acid solution, do you add the acid to water or water to acid?

14. Briefly explain why you shouldn't return unused or excess chemicals to the original container.

Chemistry Laboratory Scavenger Hunt

NAME: _____

DATE: _____

INSTRUCTOR : _____

SECTION : _____

PARTNER'S NAME: _____

Write your answers in pen and include units.

Instructions: You are required to read the questions below and complete the activity while familiarizing yourself with the laboratory. You may work with a lab partner. However, you must turn in your own completed lab report.

1. What are all the sizes of glass beakers available in the lab?

2. Where can you put your backpack, books, and other items when in the lab?

3. Describe the location of the eyewash/safety shower.

4. Where does broken glass go?

Copyright © Dr. Megan Grunert, Western Michigan University

5. Name three supplies for cleaning glassware.

6. Find the drawer for the following pieces of glassware and list how many of each are in the drawer:

    a. Evaporating dishes

    b. Watch glasses

    c. Funnels

    d. Stirring rods

    e. Test tubes

7. Where are the hot plates located?

8. When you are not using the hood, the sash (glass panel) should be in what position?

9. Where are the balances located?

10. What are balances used for?

11. What kinds of water faucets are in the lab?

12. Describe the location of the fire extinguisher.

13. Where are the gloves located?

14. Draw an Erlenmeyer flask.

15. What sizes of Erlenmeyer flasks are available in lab?

16. List at least three things that you can use to obtain the mass of a solid chemical.

17. What could be used to add small amounts of liquid to a beaker or graduated cylinder?

18. What sizes of graduated cylinders are in lab?

19. What can be used to clean up spills in the lab?

20. Where are the ring stands located?

Lab 1

Size and Scale

CHEM 1110

name _____ date _____ instructor _____

section _____ partner's name _____

Write your answers in pen and include units.

This worksheet is your lab report. It is due at the start of the next lab session.

Individual Card Arrangement (Largest = 1; Smallest = 16)

Largest

1. _____
2. _____
3. _____
4. _____
5. _____
6. _____
7. _____
8. _____
9. _____
10. _____
11. _____
12. _____
13. _____
14. _____
15. _____

Smallest 16. _____

CHEM 1110   SIZE AND SCALE

Visualization Tools:
Now that you have arranged your cards in order, decide what instrument you would need to measure each object. Write each object in the correct column.

TELESCOPE	EYE	OPTICAL MICROSCOPE	SCANNING ELECTRON MICROSCOPE	TOO SMALL TO VISUALIZE

Scale
On the connecting lines, record your answers for how many times bigger or smaller the object is compared to the Thickness of a Penny.

Moon   Human   Thickness of Penny   Cell   Atom

CHEM 1110　　　　　　　　　　　　　　　　　　　　　　　　　SIZE AND SCALE

Pre-fixes

In chemistry, we use pre-fixes in front of units to describe changes in scale. For meters, you have probably seen *centimeters, kilometers,* or *millimeters.* Next, arrange the pre-fix cards in order from **smallest** to **largest.** Then find the power of 10 that corresponds to each pre-fix. Hint: If you have ordered the cards correctly, the power of 10 should increase. You may need to use your book for this part. Finally, give an approximate size of something measured with each pre-fix. You can use items from the cards or from everyday life.

PRE-FIX NAME	SYMBOL	POWER OF 10	APPROXIMATE SIZE OF SOMETHING MEASURED
Meter	*m*	10^0	*I am 1.75 meters tall*

COMMON PREFIXES IN THE METRIC SYSTEM

PREFIX	ABBREVIATION	MEANING
Mega	M	10^6
Kilo	K	10^3
Deci	d	10^{-1}
Centi	c	10^{-2}
Milli	m	10^{-3}
Micro	μ	10^{-6}
Nano	n	10^{-9}
Pico	p	10^{-12}
Femto	f	10^{-15}

CONVERSION FACTORS IN LENGTH UNITS

1 Km = 10^3 m
1 cm = 10^{-2} m
1 mm = 10^{-3} m
1 nm = 10^{-9} m
1 Å = 10^{-10} m

Inches (in) to Feet (ft):	1 ft = 12 inches
Yard (yd) to foot (ft):	1 yd = 3 ft
Mile to feet (ft):	1 mile = 5280 ft
Inches (in) to centimeters (cm):	1 in = 2.54 cm
Meters (m) to Inches (in):	1 m = 39.37 in
Miles to kilometers (Km):	1 mile = 1.609 Km

Note: ***The diameter of the moon is: 3474.8 kilometers***

TYPES OF SCOPES

TELESCOPE: A telescope is an instrument that allows visualization of objects that are at far distances by collecting electromagnetic radiation.

OPTICAL MICROSCOPE: Optical microscopes allow visualization of objects that are on the micron scale but not much smaller. These microscopes are commonly found in your biology and other science laboratories.

TRANSMISSION ELECTRON MICROSCOPE: A transmission electron microscope (TEM) uses electron beams to pass through the sample being analyzed. The result is a two-dimensional image and the microscope allows visualization of particles as small and 1 nanometer (nm).

SCANNING ELECTRON MICROSCOPE: A scanning electron microscope (SEM) operates in a similar way as a transmission electron microscope. This type of microscope uses an electron beam to scan a sample and produce an image. An SEM produces a three-dimensional image.

SCANNING PROBE MICROSCOPE: Scanning probe microscopes (SPM) allow the imaging of particles on the nanometer to angstrom size scales. There are three types of scanning probe techniques that have become most common:

Atomic Force Microscopy (AFM): This microscope is equipped with a tip and the technique measures the force of interaction between the tip and the particle surface. This force leads to the formation of the image.

Scanning Tunneling Microscopy (STM): This microscope is also equipped with a tip. The technique measures weak electrical current that flows between the tip and the sample at a specific distance from each other. The electrical current leads to the formation of the image.

Near-Field Scanning Optical Microscopy (NSOM): In this technique, the tip scans a small light source close to the sample being analyzed and the detection of the light energy leads to the formation of the image.

Lab 2

STRC 2017

Line Spectra: Evidence for Atomic Structure

Prepared by James F Hall
University of Massachusetts Lowell

Objective

The bright-line spectra produced when excited atoms emit electromagnetic radiation has led to our modern model for atomic structure. In this experiment you will have a chance to observe first-hand some of these spectra.

Introduction

An atom that possesses excess energy is said to be in an *excited* state. When an atom in an excited state returns to its ground state, it does so my emitting excess energy as photons of electromagnetic radiation, including visible light. The internal energy states of an atom are *discrete*; these states are of fixed, specific energies that never vary. Each type of atom (each element) has unique, characteristic, discrete energy states that are different from the energy states of all other elements.

When an excited atom emits its excess energy, the energy is not emitted continuously. Atoms emit photons of only certain specific energies, which correspond exactly in energy to the changes in energy between the discrete energy states within the atom. Because atoms emit photons of only certain specific energies, the wavelengths of these emitted photons can be used to identify atoms. Typically, light from energized atoms is passed through a prism (or other device) that separates the light into its component wavelengths.

When light is separated into its component wavelengths, the pattern of different component colors produced is called a spectrum. You may have seen sunlight or light from an incandescent light bulb separated by a prism into a rainbow pattern called a continuous spectrum (see Figure 1 below).

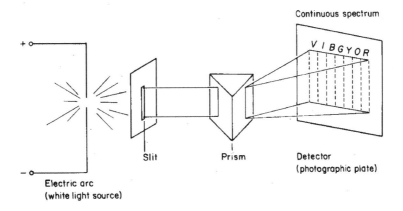

Figure 1. Continuous spectrum produced by a "white light" source

When light from a particular energized element is passed through a prism, a bright line spectrum is produced (see Figure 2 below). The colored lines occur at specific places in the spectrum (specific wavelengths) that are characteristic for each element and determined by the internal energy states in the particular atom.

© 2012, 2010 Cengage Learning. ALL RIGHTS RESERVED. No part of this work covered by the copyright herein may be reproduced, transmitted, stored, or used in any form or by any means graphic, electronic, or mechanical, including but not limited to photocopying, recording, scanning, digitizing, taping, Web distribution, information networks, or information storage and retrieval systems, except as permitted under Section 107 or 108 of the 1976 United States Copyright Act, without the prior written permission of the publisher.

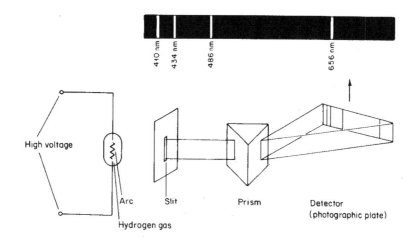

Figure 2. The spectrum of hydrogen
When light from a hydrogen lamp is passed through a prism, only certain bright lines of characteristic wavelength are observed.

Line spectra, and more modern derivations, are routinely used in chemical analysis laboratories to detect the presence of a particular element in a sample. As an example, if an analyst believes that an unknown sample contains iron, he or she can generate a bright line spectrum for a known sample of iron, and then compare the spectrum to that of the unknown sample. If the characteristic spectral lines of iron are present in the spectrum of the unknown sample, then the sample must, indeed, contain iron. Modern instruments can also quantify spectral data, and can determine how much of a particular element is present in the sample.

In this experiment, you will use a Direct-Reading spectroscope to view the bright line spectra of several elements. The spectroscope consists of a diffraction grating with an internal scale that shows wavelength in nanometers. The grating, which has the same effect on light as a prism, has 600 lines per millimeter. When aligned with the light source, the slit allows the viewer to see the detected spectrum on a back-lit scale, which reads from 380 to 720 nm.

Two sources will be used to generate spectra. First of all, your instructor will set up one or more gas discharge tubes for you to view. A gas discharge tube is typically a long, narrow bulb filled with a particular gaseous element. The tube is fitted with metal electrodes at each end. When a high voltage is applied to the metal electrodes, the atoms of gaseous element absorb, and then reemit, energy, partially as visible light. When the light from a gas discharge tube is viewed through the spectroscope, the bright line spectrum of the gaseous element can be seen. The second method of producing spectra you will perform yourself: samples of various metal-ion salts will be sprayed into a burner flame, and the light emitted by the energized metal ions will be viewed through the spectroscope.

Apparatus/Reagents Required

- spectroscope
- gas discharge tubes and power supply
- spray bottles containing 0.1 M solutions of the salts NaCl, KCl, $SrCl_2$, $CaCl_2$ and $CuCl_2$.
- plastic sheeting to cover the lab bench area

Safety Precautions	• Safety eyewear approved by your institution must be worn at all times while you are in the laboratory, whether or not you are working on an experiment. • Gas discharge tubes also emit other wavelengths of electromagnetic radiation, in particular, ultraviolet radiation, which may be damaging to the eyes. Your safety goggles will absorb most ultraviolet radiation, so keep them *on* while observing the spectra. • The power supply for the gas discharge tube uses very high voltages: *do not handle or attempt to adjust the power supply yourself.* • The salts used in the flame tests may be toxic if ingested. Wash after handling. Wash down your lab bench to remove residues of the salts.

Procedure

1. The spectroscope

Various types of spectroscopes may be available for your use. Examine the spectroscope you will be using for this experiment. One end of the long tube should have a round viewing hole fitted with a small piece of plastic diffraction grating. The other end of the spectroscope should have a narrow slit. Try looking at the lighting fixtures in the laboratory through the spectroscope: if your laboratory has fluorescent tube lighting, align the slit of the spectroscope parallel to the light tubes. You should see a rainbow pattern (continuous spectrum) off to either side of the slit. If you do not see such a rainbow pattern, try rotating the eyepiece of the spectroscope (it contains the diffraction grating). If you do not see a rainbow spectrum at this point, consult with the instructor.

2. Gas Discharge Tubes

Your instructor will set up a high voltage power supply and one or more gas discharge tubes. *You should not handle this equipment yourself:* the power supply generates more than 5,000 volts and is too dangerous to be handled by students.

The first gas discharge tube your instructor will demonstrate will be for hydrogen. After the room lights are dimmed, and the hydrogen tube has been turned on, view the hydrogen tube through the spectroscope. Align the slit of the spectroscope parallel to the length of the hydrogen tube.

Try to view from no more than 4 to 5 cm away from the hydrogen tube so that the spectrum is easier to see.

Record the color of the lines you see, as well as their relative order from left to right. Nearly all students will see three bright lines (red, blue-green, blue), while some students with keen eyesight may see a fourth (violet) line. Compare the spectrum you see with that shown in your textbook.

Next, your instructor will set up a gas discharge tube containing water vapor. Since water contains the elements hydrogen and oxygen, how do you expect the spectrum of water to compare to the individual spectrum of hydrogen? View the spectrum of water and record the color of the lines you see as well as their relative order from left to right.

Finally, depending on what's available in your laboratory, your instructor may set up additional gas discharge tubes for you to view. In each case, record the color and relative positions of the bright lines in the spectra.

Name: _____ Section: _____

Lab Instructor: _____ Date: _____

Pre-Laboratory Questions

1. Use your textbook to write definitions or explanations for each of the following terms or concepts.

    a. the wavelength of electromagnetic radiation

    _____
    _____
    _____
    _____
    _____

    b. the frequency of electromagnetic radiation

    _____
    _____
    _____
    _____
    _____

    c. the speed of electromagnetic radiation

    _____
    _____
    _____
    _____
    _____

    d. the ground state of an atom

    _____
    _____
    _____
    _____
    _____

e. why only certain wavelengths of light are emitted by excited atoms of a given element

f. what it means to say that energy levels of an atom are quantized

2. Why should you not attempt to manipulate the power supply used to illuminate the gas discharge tubes in this experiment?

3. Why must safety goggles be worn while viewing the line spectra from the gas discharge tubes?

Line Spectra: Evidence for Atomic Structure

Name: _____ Section: _____

Lab Instructor: _____ Date: _____

Results/Observations

1. Observation of fluorescent light "rainbow" spectrum. List the colors you observed, in order from left to right, in the spectrum.

   _____
   _____
   _____
   _____

2. Gas Discharge Tubes

   List the colors of the lines you observed in each spectrum, from left to right, as well as any other observations of the spectrum.

   Hydrogen

   _____ _____ _____ _____ _____ _____ _____

   Water vapor

   _____ _____ _____ _____ _____ _____ _____

   Other element(s)

   _____ _____ _____ _____ _____ _____ _____

   _____ _____ _____ _____ _____ _____ _____

   Other observations on the spectra:

   _____
   _____
   _____

Questions

1. Some of the spectra may have contained regions that were "continuous" (that is, rather than sharp, bright lines, some regions may have shown a continuous band of color or colors). What might be responsible for such a region of continuity?

2. How did the spectrum of water vapor compare to the individual spectra of hydrogen? Why might this be expected?

Lab 3
Molecular Representations: Lewis Structures, 3D Shapes and Physical Models

Prepared by Kelley M. Current and Brianna N. Hyder,
© 2014 Western Michigan University

Objectives

Given a molecular formula, students will be able to indicate a compound's: (a) Lewis Structure, (b) Electronic Geometry, (c) Molecular Geometry, and (d) Hybridization.

Why we draw Molecular Representations

Chemists frequently deal with objects (elements and compounds) so small that the naked eye cannot directly observe them. These objects are said to exist on the **micro-scale**. Chemists are interested in understanding and manipulating elements and compounds using various laboratory techniques. These laboratory techniques exist on the observable **macro-scale**. Throughout this course you will carry out a series of laboratory experiments. These experiments have been carefully designed so that the macro-scale manipulations called for in the procedure (weighing, heating, mixing, dissolving, etc.) affect objects on the micro-scale and yield the desired chemical products or data. Because chemists cannot directly manipulate objects at the micro-scale, macro-scale manipulations are applied for the same effect.

Since chemists are interested in objects which exist at the micro-scale and which cannot be directly observed, they have developed a set of standard terms and rules that are used to depict chemical compounds. In this lab you will be presented with these terms and rules.

Lewis Structures

Given a molecular formula (a formula indicating the types and proportions of atoms present in a compound), you will be asked to determine how the atoms in that molecule are connected and arranged in three dimensional (3D) space. The atoms of a compound are held together by **chemical bonds** (the attractive force between adjacent atoms which results from the placement of electrons).

In general, metals combine with non-metals and the resulting bonds are ionic. However, it is important to take into consideration that the type of bonds that form depend on the differences in electronegativity. Small differences in electronegativity (0 – 0.4) result in non-polar covalent bonds. If the difference in electronegativity is slightly higher (0.4 – 1.7), polar covalent bonds will form, for example in the case of Be and Cl. If the chemical bonds are **covalent**, the electrons are shared between two atoms (both non-metals) within a compound and the resultant bond is depicted as a line:

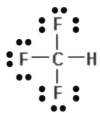

Figure 1: In trifluoromethane, a carbon shares electrons (is covalently bonded) with three fluorine atoms and one hydrogen atom. The dots surrounding the fluorine atoms represent lone pairs of electrons surrounding the fluorine atoms.

When atoms have differences in electronegativity higher than 1.7, then ionic bonds will form. If the chemical bonds are **ionic**, the electrons completely leave one atom (the metal) in favor of the other atom (the non-metal). An electrostatic charge holds these atoms together. The resultant bond is depicted using brackets and charges in Figure 2.

$$[:\ddot{C}l:]^{-1} \quad [Mg]^{+2} \quad [:\ddot{C}l:]^{-1}$$

Figure 2: A series of magnesium and chloride ions are held together by ionic bonds. Both of magnesium's valence electrons are transferred to individual chloride ions.

Figures 1 and 2 both represent Lewis Structures; they indicate the order in which atoms are connected and arranged around a central atom. Lewis Structures do not need to depict the 3D arrangement of atoms in space. The Lewis Structures shown in Figures 1 and 2 also adhere to the octet and duet rules.

The **octet rule** states that: All atoms with atomic numbers of 20 or less, combine to have eight electrons in their outermost (valence) shell. An immediate exception to the octet rule is the duet rule. The **duet rule** states that: Hydrogen and helium atoms do not require eight valence electrons; only two electrons are required to fill their valence shells. The general procedure (outlined in Figure 3) can be used to draw Lewis Structure which adhere to the octet/duet rules.

1. Find the total number of valence electrons for all atoms in a molecule
    a. Add electrons for each negative charge in an anion
    b. Subtract electrons for each positive charge in a cation
    c. Divide the sum by 2 to determine the number of electron pairs
2. Identify central atom (never hydrogen) and connect the atoms; draw lines to represent bonding pairs
    a. Hydrogen and halogens usually form only one bond
    b. Elements in the second row are usually predictable
    c. Elements near the boundary between metals and non-metals (such as Be and B) may have less than an octet; never draw multiple bonds for these elements
    d. Elements in the third row and below may have exceptions to the octet rule
3. Subtract the number of valence electrons used for bonding from the total number calculated in step 1 and assign to terminal atoms to fulfill octets
4. Place remaining unassigned electrons on the central atom as lone pairs
Octets may need to be achieved by multiple bonds if there are unassigned electrons left
5. Determine the formal charge for all atoms in the Lewis Structure using the following formula: Formal Charge = # of valence electrons - # of nonbonding electrons - # of bonds

Figure 3: Procedure and considerations for depicting Lewis Structures.

Once a Lewis Structure has been determined, it can be used to predict the structure's 3D arrangement by referencing Figure 4.

Summary of Electronic and Molecular Geometries

Number of Substituents	Number of Bonds	Number of Lone Pairs	Electronic Geometry	Molecular Geometry	Bond Angles	Hybridization
2	2	0	Linear	Linear	180	sp
3	3	0	Trigonal planar	Trigonal planar	120	sp^2
3	2	1	Trigonal planar	Angular (or bent)	<120	sp^2
4	4	0	Tetrahedral	Tetrahedral	109.5	sp^3
4	3	1	Tetrahedral	Trigonal pyramidal	107	sp^3
4	2	2	Tetrahedral	Angular (or bent)	104.5	sp^3
4	1	3	Tetrahedral	Linear		sp^3
5	5	0	Trigonal bipyradmidal	Trigonal bipyradmidal	90, 120, 180	sp^3d
5	4	1	Trigonal bipyradmidal	See-saw		sp^3d
5	3	2	Trigonal bipyradmidal	T-shape		sp^3d
5	2	3	Trigonal bipyradmidal	Linear		sp^3d
6	6	0	Octahedral	Octahedral	90, 180	sp^3d^2
6	5	1	Octahedral	Square pyramidal		sp^3d^2
6	4	2	Octahedral	Square planar		sp^3d^2

Figure 4: Use the number of substituents to determine an atom's geometry (both electric and geometric). Substituents reflect the central atom's valence number and include the number of bonds of shared electrons (double and triple bonds are counted as one connection) and lone pairs.

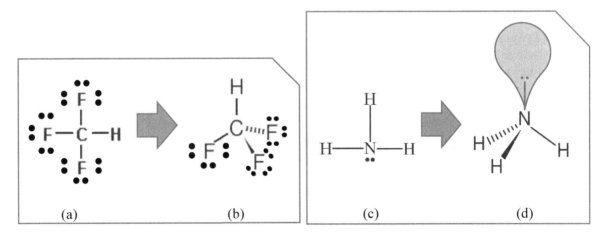

Figure 5: Lewis Structures converted to 3D depictions using information from Figure 4.

Input	Name	Trifluoromethane	Ammonia
		Substituents: 4	Substituents: 4
		Bonds: 4	Bonds: 3
		Lone Pairs: 0	Lone Pairs: 1
Output	Electronic Geometry	Tetrahedral	Tetrahedral
	Molecular Geometry	Tetrahedral	Trigonal Pyramidal
	Bond Angles	109.5	107
	Hybridization	sp^3	sp^3

Figure 6: By counting the number of substituents, bonds, and lone pairs a compound has, Figure 5 can be used to determine the electronic and molecular geometry of a compound. Bond angles and hybridization are also indicated.

VSEPR Theory

Bond angles indicated in Figures 4, 5, and 6 are predictions, which are the result of a collection of rules, referred to as **Valence Shell Electron Pair Repulsion (VSEPR) Theory**. VSEPR theory allows for the prediction of bond angles based on Lewis Structures. The VSEPR rules state that the valence electron groups (bonds or lone pairs) surrounding an atom repel each other. Recall that electrons are negatively charged, and that like charges repel. For every atom of every compound, electron groups adopt an arrangement that minimizes this repulsive force. *The arrangement that most successfully minimizes the repulsion between electron groups (bonds or lone pairs) dictates a compound's molecular geometry.*

Notice that trifluoromethane and ammonia have different molecular geometries and bond angles. Though both compounds have four substituents, ammonia has a lone pair of electrons that trifluoromethane does not. VSEPR Theory states that *the repulsion caused by a lone pair is greater than the repulsion caused by a pair of bonding electrons*. Because ammonia's lone pair is more repulsive than the N-H bonding electrons, the angles between N-H bonds are forced together, decreasing their bond angles (from 109.5 degrees to 107 degrees).

Physical Properties

Once determined, molecular geometry can be used to make predictions about a compound's physical properties. Trifluoromethane is comprised of four bonds: three C-F and one C-H. C-F bonds are said to be polar-covalent bonds. **Polar-covalent bonds** are those in which the bonding electrons are more attracted to one atom of the bonding pair than the other. **Nonpolar-covalent bonds** are those in which bonding electrons are equally attracted to each of the bonded atoms. C-F bonds are polar-covalent bonds, while C-H bonds are nonpolar covalent bonds. C and H have very similar electronegativities, or a similar ability to attract electrons. C and F on the other hand have very different electronegativities. F is more able to attract electrons than C and the electrons in the C-F bond are more frequently found with F than with C.

The polarity of a compound can be predicted using 3D Lewis Structures in combination with the nature of the bonds which comprise the structure. In both trifluoromethane and ammonia, there are polar bonds. C-F and N-H are polar bonds. Trifluoromethane also has a non-polar C-H bond. Both trifluoromethane and ammonia are polar compounds, because they lack symmetry.

Hybrid Orbitals

There are other methods (aside from VSEPR Theory) which allow predictions of molecular geometry and physical properties. **Valence Bond Theory** describes how atomic orbitals (spaces electrons occupy around an atomic core) are mixed to form hybrid orbitals (spaces electrons occupy between bonded atoms).

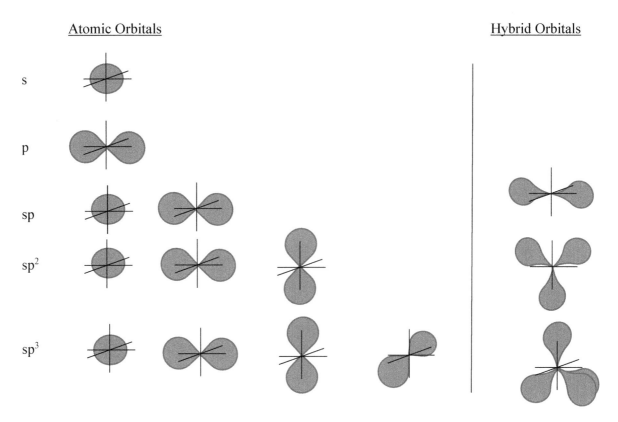

Figure 7: The overlap of hybridized orbitals between bonded atoms is another way to conceptualize bonding and predict the 3D structure of a compound.

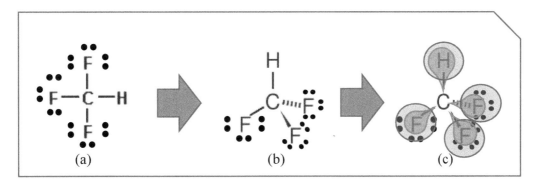

Figure 8: Structures (a) & (b) both represent Lewis Structures previously shown. (c) Depiction of the hybridized atomic orbitals of the central carbon atom (sp^3) overlapping the atomic orbital of fluorine (s).

Using Valence Bond Theory to depict the overlap of atomic and molecular orbitals provides a better means of visualizing where electrons exist within a given molecule and between bonded atoms.

In today's laboratory exercise you will be investigating the following:
(1) Lewis Structure: Diagrams showing the arrangement in which the atoms of a compound are connected.
(2) Electronic Geometry: Diagram showing the 3D arrangement of atoms and electron lone pairs in space. In these diagrams the lone pairs of electrons **are** considered when determining the geometry of a given compound.
(3) Molecular Geometry: Diagram showing the 3D arrangement of atoms and electron lone pairs in space. In these diagrams the lone pairs of electrons **are not** considered when determining the geometry of a given compound.
(4) Hybridization: Explanation of how atomic orbitals are mixed to form hybrid orbitals in three dimensions.

Provided below is a sample of a worked exercise analogous to what you will complete today.

	Lewis Structure	3D Shape with bond angles	Polar?
3. H_2S Circle one: ionic or (covalent) Valence electrons: 2(1)+6=8 Electron pairs: 8/2 = 4 Molecular Geometry: angular	H–S̈–H	(bent structure with H, S, H)	yes Hybridization sp^3

PROCEDURE

Part A: Identification of Molecular Geometries and Hybridization

It is important to understand the differences in molecular geometry. You will be given a collection of cards with different geometric shapes. You will need to record in the report form the letter of the card, the name of the molecular geometry, and the hybridization.

Furthermore, you will use these cards to understand hybridization. For parts B-D, you will use the codes on the cards to indicate the hybridization of the molecules you build.

Part B: Determination of Lewis Structure, Molecular Geometry, Bond Angles, and Polarity

In this activity, you will be working with a partner and other groups to draw Lewis Structures and make physical models of specific molecules. Your instructor will assign you a series of molecules from the chart on the report form. You will need to follow the steps of the procedure for depicting Lewis Structures on Figure 3 to properly determine the Lewis Structure and build the geometric shape with a ball and stick modeling kit. If protractors are available, you may use them to measure bond angles. You need to draw the 3D shape from your model, including bond angles. Record whether the compound is polar and record the letter code from the hybridization cards. Then find other groups in the lab to confirm your model. Resolve any disagreements with your peers. Once you are confident with your construction, share the Lewis Structure diagram and your model with the rest of the class.

Advanced – your TA may ask you to draw the hybridized orbitals of the molecules as well.

Part C: Compounds with Multiple Bonds

Provide two or more potential connectivities for each compound in this section. Calculate the formal charge for each element in each compound. Based on formal charge minimization, select the most appropriate connectivity. Complete the Lewis Structures by assuring that all atoms adhere to the octet or duet rule.

Part D: Compounds with Charges

You will determine the structure of the four molecules. Remember to show proper notation of the charge.

Advanced – your TA may ask you to show the resonance structures of a molecule as well.

Molecular Representations

Name _____ Date _____ Instructor _____

Section _____ Partner's Name _____

Write your answers in pen

Pre-laboratory Questions

1. Which compound is formed by a covalent bond?
   a. $BaCl_2$ b. $MgCl_2$ c. Cl_2 d. KI   e. $CuCl_2$

2. Indicate the number of valence electrons for the ions in the compounds below:
   a) SO_4^{2-}
   b) I^-
   c) NO_2^+
   d) NH_4^+
   e) H_3O^+
   f) CH_4

3. Use your own words to define the octet rule and describe an immediate exception.

4. Which has an incomplete octet in its Lewis structure?
   a. NO   b. SO_2 c. ICl   d. CO_2 e. Cl_2

5. Consider the molecule BF_3

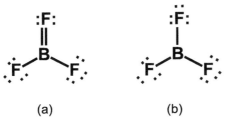

(a)   (b)

What are the formal charges of the atoms in Lewis structure (a) and (b)?

6. Which compound has a different hybridization than the others?
    a) $COCl_2$ b) BF_3 c) NH_3 d) H_2CO

7. Identify the following compounds as the polar or non-polar and give an explanation for that particular behavior: NH_3, CO_2, BF_3, SO_2

8. Which one has the lowest boiling point (b.p.) based on their polarities?
    a) H_2O b) CH_4 c) $SeCl_2$ d) $CHCl_3$

Name _____ Date _____ Instructor _____
Section _____ Partner's Name _____ Group # _____
Write your answers in pen.

Molecular Representations

REPORT FORM

Part A (10 points) – Identification of Molecular Geometries and Hybridization

Collect the set of cards for this activity. Record the letter of each card, write the molecular geometry observed and include the hybridization. Refer to these answers for the remaining exercises.

Card Letter	Molecular Geometry	Hybridization

Part B (48 points) - Consult the steps of Figure 3 to draw the correct Lewis structure and build the geometric shape with a ball and stick modeling kit for your assigned molecules. Find other groups in the lab to confirm your model. After reaching a consensus, share the Lewis structure diagram and model with the rest of the class. Record and check the correct answers for the remaining structures.

Molecule	Lewis Structure	3D Shape with bond angles	Polar? Hybridization
1. $BaCl_2$ Circle one: ionic or covalent Valence electrons: Electron pairs: Molecular Geometry:			
2. KCl Circle one: ionic or covalent Valence electrons: Electron pairs: Molecular Geometry:			
3. H_2S Circle one: ionic or covalent Valence electrons: Electron pairs: Molecular Geometry:			
4. NH_3 Circle one: ionic or covalent Valence electrons: Electron pairs: Molecular Geometry:			
5. $BeCl_2$ Circle one: ionic or covalent Valence electrons: Electron pairs: Molecular Geometry:			
6. BH_3 Circle one: ionic or covalent Valence electrons: Electron pairs: Molecular Geometry:			

Molecule	Lewis Structure	3D Shape with bond angles	Polar? Hybridization
7. H₂O Circle one: ionic or covalent Valence electrons: Electron pairs: Molecular Geometry:			
8. CH₄ Circle one: ionic or covalent Valence electrons: Electron pairs: Molecular Geometry:	Lewis Structure	3D Shape with bond angles	Polar? Hybridization
9. SeCl₂ Circle one: ionic or covalent Valence electrons: Electron pairs: Molecular Geometry:	Lewis Structure	3D Shape with bond angles	Polar? Hybridization
10. CO₂ Circle one: ionic or covalent Valence electrons: Electron pairs: Molecular Geometry:	Lewis Structure	3D Shape with bond angles	Polar? Hybridization
11. BrF₃ Circle one: ionic or covalent Valence electrons: Electron pairs: Molecular Geometry:	Lewis Structure	3D Shape with bond angles	Polar? Hybridization
12. XeF₄ Circle one: ionic or covalent Valence electrons: Electron pairs: Molecular Geometry:	Lewis Structure	3D Shape with bond angles	Polar? Hybridization

Practice – Show the calculations for the formal charge of each atom in carbon dioxide:
(Atom's number of valence electrons minus number of electrons assigned to it)

Part C (16 points) - Predict two or more structures for the following compounds that may have multiple bonds. Based on formal charge minimization, select the appropriate structure and identify the molecular geometry and hybridization of one of the central atoms.

	Lewis Structure with bond angles	Molecular Geometry (circle atom in diagram)	Hybridization
C_2H_6 Compound name:			
C_2H_4 Compound name:			
C_2H_2 Compound name:			
HNO_3 *Show formal charges Compound name:			

Advanced – Show the resonance structures for the carbonate ion, CO_3^{2-}:

Part D (16 points) - Predict the following structures, including charges, and determine the hybridization.

	Lewis Structure with charges	Molecular Geometry	Hybridization
OH^-			
N_3^-	Lewis Structure with charges	Molecular Geometry	Hybridization
NO_3^-	Lewis Structure with charges	Molecular Geometry	Hybridization
NH_4^+	Lewis Structure with charges	Molecular Geometry	Hybridization

*Show formal charges

POST-LABORATORY (10 Points)

Now that you have practiced finding different representations of compounds, you will need to take what you've learned to write a report on the relationship of molecular structure and physical properties. From the list of compounds below, you will need to identify the following:
1) Chemical formula
2) Chemical bond – ionic or covalent
3) Lewis structure
4) Molecular geometry and hybridization
5) Polarity
6) Commercial product or real-life practical use
7) Physical properties: color, freezing or boiling point, melting point, density, molecular weight, etc.

List of Compounds:
Borax	Butane	Calcium carbide
Carbon monoxide	Fructose	Silica
Toluene		

If you want to do a different compound, ask your instructor for approval.

Give an explanation as to why your compound exhibits these properties and be sure to provide proper citation. Consult your instructor for the format of the report.

Example: sodium chloride

> Sodium chloride has the chemical formula NaCl and its molecular weight is 58.44 grams per mole. The chemical bond is ionic and the ions tend to form a crystalline structure appearing as white solid and having a melting point of 801°C (Lide). It is commonly referred to as salt and it is soluble in water as both are polar. It is sprinkled on roads and sidewalks during the winter to lower the freezing point of water and therefore prevent formation of ice, due to colligative properties of the mixture.
>
> 1. Lide, David R. "Physical Constants of Inorganic Compounds." *CRC Handbook of Chemistry and Physics 1913-1995.* 75th Ed. Boca Raton, FL: CRC Press, Inc. 1994; page 4-98.
>
> Here is the structure of the salt crystal:

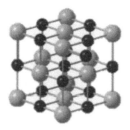

Lab 4

Identification of a Compound by Mass Relationships

Prepared by Emil Slowinski; Wayne C. Wolsey; William L. Masterton, Macalester College

PROP 2122

When chemical reactions occur, there is a relationship between the masses of the reactants and products that follows directly from the balanced equation for the reaction and the molar masses of the species that are involved. In this experiment we will use this relationship to identify an unknown substance.

Your unknown will be one of the following compounds, all of which are salts:

$$NaHCO_3 \quad Na_2CO_3 \quad KHCO_3 \quad K_2CO_3$$

In the first part of the experiment you will be heating a weighed sample of your compound in a evaporating dish. If your sample is a carbonate, there will be no chemical reaction that occurs, but any small amount of adsorbed water will be driven off. If your sample is a hydrogen carbonate, it will decompose by the following reaction, using $NaHCO_3$ as the example:

$$2\ NaHCO_3(s) \rightarrow Na_2CO_3(s) + H_2O(g) + CO_2(g) \tag{1}$$

In this case there will be an appreciable decrease in mass, since some of the products will be driven off as gases. If such a mass decrease occurs, you can be sure that your sample is a hydrogen carbonate.

In the second part of the experiment, we will treat the solid carbonate in the evaporating dish with HCl, hydrochloric acid (HCl). (Please note that HCl is corrosive and must be handled with care). There will be considerable effervescence as CO_2 gas is evolved; the reaction that occurs is, using Na_2CO_3 as our example:

$$Na_2CO_3(s) + 2\ H^+(aq) + 2\ Cl^-(aq) \rightarrow 2\ NaCl(s) + H_2O(l) + CO_2(g) \tag{2}$$

(Since HCl in solution exists as ions, we write the equation in terms of ions.) We then heat the evaporating dish strongly to drive off any excess HCl and any water that is present, obtaining pure, dry, solid NaCl as our product.

To identify your unknown, you will need to find the molar masses of the possible reactants and final products. For each of the possible unknowns there will be a different relationship between the mass of the original sample and the mass of the chloride salt that is produced in Reaction 2. If you know your sample is a carbonate, you need only be concerned with the mass relationships in Reaction 2, and should use as the original mass of your unknown the mass of the carbonate after it has been heated. If you have a hydrogen carbonate, the overall reaction your sample undergoes will be the sum of Reactions 1 and 2.

From your experimental data you will be able to calculate the change in mass that occurred when you formed the chloride from the hydrogen carbonate or the carbonate that you started with. That difference divided by the mass of the original salt will be different for each of the possible starting compounds and will not vary with the mass of the sample. Let's call that quantity Q.

$$Q = (\text{mass of chloride} - \text{original mass})/\text{original mass} \qquad (3)$$

Your calculation of the theoretical values of Q for each of the possible compounds should allow you to determine the identity of your unknown. Since you will already know whether your compound is a carbonate or a hydrogen carbonate, you should then need only work out which of the two possible compounds of that type yours may be.

Experimental Procedure

Obtain an unknown from your instructor.

Clean your evaporating dish and a watch glass by rinsing them with distilled water and then drying them with a towel. Place the evaporating dish with its cover slightly ajar on a hot plate. Heat gently on a hot plate for a minute or two and then strongly for two more minutes. Allow the evaporating dish to cool to room temperature (it will take about 10 minutes; it must cool completely, or your mass measurement will be artificially lowered by convection currents).

Weigh the evaporating dish accurately on an analytical balance. Record the mass on the Data page. With a spatula, transfer about 0.5 g of the unknown to the evaporating dish. Weigh the evaporating dish and the sample of unknown on the balance. Record the mass.

Heat the evaporating dish, gently and intermittently, for a few minutes. Gradually increase the hot plate temperature, to the point where the bottom of the evaporating dish is hot. Heat for 10 minutes. Allow the evaporating dish to cool for 10 minutes, and then weigh it, with its cover and contents on the analytical balance, recording the mass.

At this point the sample in the evaporating dish is a dry carbonate, since the heating process will convert any hydrogen carbonate to carbonate.

Add about 25 drops of 6 M HCl, a drop at a time, to the sample. As you add each drop, you will probably observe effervescence as CO_2 is produced. Let the action subside before adding the next drop, to keep the effervescence confined to the lower part of the evaporating dish. We do not want the product to foam up over the edge! When you have added all of the HCl, the effervescence should have ceased, and the solid should be completely dissolved. Heat the evaporating dish gently for brief periods to complete the solution process. If all of the solid is not dissolved, add 6 more drops of 6 M HCl and warm gently.

Place the cover on the evaporating dish in an off-center position, to allow water to escape during the next heating operation. Heat the evaporating dish, gently and intermittently, for about 10 minutes, to slowly evaporate the water and excess HCl. If you heat too strongly, spattering will occur and you may lose some sample.

Allow the evaporating dish to cool for 10 minutes. Weigh it, with its contents, on the analytical balance and record, the mass.

DISPOSAL OF REACTION PRODUCTS. Dispose of the sample as directed to you by your instructor.

Pre-Lab Questions

_____   _____ _____
name   date   instructor

_____   _____
section   partner's Name

Write your answers in pen and include units.

1. In this experiment, two of the unknowns have the common names washing soda and baking soda. Which are these?

2. Potassium hydrogen carbonate is used in some dry chemical fire extinguishers. With the aid of an equation describe what happens if you heat $KHCO_{3(s)}$.

3. Why must the evaporating dish be allowed to cool completely after heating?

4. When measuring the unknown sample, is it necessary to take *exactly* 0.5 g of the compound?

5. When heating, why should you heat the evaporating dish gently and intermittently?

6. Which one is not correct:
   (a) A mass decrease after heating is the proof of having a hydrogen carbonate compound as the reactant.
   (b) A considerable effervescence when HCl is added to a hydrogen carbonate is the result of H_2O being evolved.
   (c) A hydrogen carbonate sample can be easily decomposed by heating.

_____   _____   _____
name   date   instructor

_____   _____
section   partner's Name

Write your answers in pen and include units.

Report Form

Identification of a Compound by Mass Relationships

Unknown Number: _____

Atomic Masses: Na_____g   K_____g   H_____g

C_____g   O_____g   Cl_____g

Molar Masses: $NaHCO_3$_____g/mol Na_2CO_3_____g/mol   NaCl_____g/mol

$KHCO_3$_____g/mol K_2CO_3_____g/mol   KCl_____g/mol

DATA

Mass of evaporating dish and unknown   _____g

Mass of evaporating dish   _____g

Mass of unknown   _____g
(show calculation)

Mass of evaporating dish and unknown after heating   _____g

Loss of mass of sample   _____g
(show calculation)

Mass of evaporating dish and solid chloride   _____g

Mass of solid chloride   _____g
(show calculation)

DETERMINE THE FOLLOWING
BASED ON MEASUREMENTS RECORDED

Circle the possible identity of your unknown:

    carbonate   (**hydrogen carbonate**)

Change in mass when original compound was converted to a chloride   −.007 g
(show calculation)

Q = Change in mass / original mass   −.01866
(show calculation)

Theoretical values of Q, as obtained by Equation (3)

    Possible *sodium* compound $NaHCO_3$ | Show calculations:

    1 mole of compound → ____1____ moles NaCl

    __84.01__ g of compound → __58.44__ g NaCl

    Change in mass __−25.61__ g Q = __−.304__

    Possible *potassium* compound __$KHCO_3$__ | Show calculations:

    1 mole of compound → ____1____ moles KCl

    __100.1__ g of compound → __74.55__ g KCl

    Change in mass __−25.5__ g Q = __−.2552__

Identity of unknown $NaHCO_3$ −0.327

What sources of error carried in this experiment could have led to poor results?

name _____ date _____ instructor _____

section _____ partner's Name _____

Write your answers in pen and include units.

Post Lab Questions: Identification of a Compound by Mass Relationships

1. A student attempted to identify an unknown compound by the method used in this experiment. She found that when she heated a sample weighing 0.4862 g, the mass barely changed, dropping to 0.4855 g. When the product was converted to a chloride, the mass went up, to 0.5247 g.

   a. Is the sample a carbonate?   Yes / no (Circle one)

      Please provide your reasoning below.

   b. What are the two compounds that might be in the unknown?

      _____ or _____

   c. Write the balanced chemical equation for the overall reaction that occurs when each of these two original compounds is converted to a chloride. If the compound is a hydrogen carbonate, use the sum of Reactions 1 and 2. If the sample is a carbonate, use Reaction 2. Write the equation for a sodium salt and then for a potassium salt.

   d. How many moles of the chloride salt would be produced from one mole of original compound?

      _____

   e. How many grams of the chloride salt would be produced from one molar mass of original compound?

      Molar masses: $NaHCO_3$ _____ g Na_2CO_3 _____ g $NaCl$ _____ g

      $KHCO_3$ _____ g K_2CO_3 _____ g KCl _____ g

      If a sodium salt, _____ g original compound → _____ g chloride

      If a potassium salt, _____ g original compound → _____ g chloride

   f. What is the theoretical value of Q, as found by Equation 3,

      if she has the Na salt? _____ if she has the K salt? _____

   g. What was the observed value of Q?   _____

   h. Which compound did she have as an unknown?   _____

Lab 5
Determination of a Chemical Formula

Prepared by Emil Slowinski; Wayne C. Wolsey; William L. Masterton, Macalester College

When atoms of one element combine with those of another, the combining ratio is typically an integer or a simple fraction; 1:2, 1:1, 2:1, and 2:3 are ratios one might encounter. The simplest formula of a compound expresses that atom ratio. Some substances with the ratios we listed include $CaCl_2$, KBr, Ag_2O, and Fe_2O_3. When more than two elements are present in a compound, the formula still indicates the atom ratio. Thus the substance with the formula Na_2SO_4 indicates that the sodium, sulfur, and oxygen atoms occur in that compound in the ratio 2:1:4. Many compounds have more complex formulas than those we have noted, but the same principles apply.

To find the formula of a compound we need to find the mass of each of the elements in a weighed sample of that compound. For example, if we resolved a sample of the compound NaOH weighing 40 grams into its elements, we would find that we obtained just about 23 grams of sodium, 16 grams of oxygen, and 1 gram of hydrogen. Since the atomic mass scale tells us that sodium atoms have a relative mass of 23, oxygen atoms a relative mass of 16, and hydrogen atoms a relative mass of just about 1, we would conclude that the sample of NaOH contained equal numbers of Na, O, and H atoms. Since that is the case, the atom ratio Na:O:H is 1:1:1, and so the simplest formula is NaOH. In terms of moles, we can say that that one mole of NaOH, 40 grams, contains one mole of Na, 23 grams, one mole of O, 16 grams, and one mole of H, 1 gram, where we define the molar mass to be that mass in grams equal numerically to the sum of the atomic masses in an element or a compound. From this kind of argument we can conclude that the atom ratio in a compound is equal to the mole ratio. We get the mole ratio from chemical analysis, and from that the formula of the compound.

In this experiment we will use these principles to find the formula of the compound with the general formula $Cu_xCl_y \cdot zH_2O$, where the x, y, and z are integers which, when known, establish the formula of the compound. (In expressing the formula of a compound like this one, where water molecules remain intact within the compound, we retain the formula of H_2O in the formula of the compound.)

The compound we will study, which is called copper chloride hydrate, turns out to be ideal for one's first venture into formula determination. It is stable, can be obtained in pure form, has a characteristic blue-green color which changes as the compound is changed chemically, and is relatively easy to decompose into the elements and water. In the experiment we will first drive out the water, which is called the water of hydration, from an accurately weighed sample of the compound. This occurs if we gently heat the sample to a little over 100°C. As the water is driven out, the color of the sample changes from blue-green to a tan-brown color similar to that of tobacco. The compound formed is anhydrous ("no water") copper chloride. If we subtract its mass from that of the hydrate, we can determine the mass of the water that was driven off, and, using the molar mass of water, find the number of moles of H_2O that were in the sample.

In the next step we need to find either the mass of copper or the mass of chlorine in the anhydrous sample we have prepared. It turns out to be much easier to determine the mass of the copper, and find the mass of chlorine by difference. We do this by dissolving the anhydrous sample in water, which gives us a green solution containing copper and chloride ions. To that solution we add some aluminum metal wire. Aluminum is what we call an active metal; in contact with a solution containing copper ions, the aluminum metal will react chemically with those ions, converting them to copper metal. The aluminum is said to reduce the copper ions to the metal, and is itself oxidized. The copper metal appears on the wire as the reaction proceeds, and has its typical red-orange color. When the reaction is complete, we remove the excess Al, separate the copper from the solution, and weigh the dried metal. From its mass we can calculate the number of moles of copper in the sample. We find the mass of chlorine by subtracting the mass of copper from that of the anhydrous copper chloride, and from that value determine the number of moles of chlorine. The mole ratio for Cu:Cl:H_2O gives us the formula of the compound.

Determination of a Chemical Formula

Experimental Procedure

Weigh a clean, dry evaporating dish, without a cover, accurately on an analytical balance. Place about 1 gram of the unknown hydrated copper chloride in the evaporating dish. With your spatula, break up any sizeable crystal particles by pressing them against the wall of the evaporating dish. Then weigh the evaporating dish and its contents accurately. Enter your results on the Report form.

Place the evaporating dish on the table gently. Turn on your hot plate. Gently heat the evaporating dish. Do not overheat the sample. As the sample warms, you will see that the green crystals begin to change to brown around the edges. Continue gentle heating, slowly converting all of the hydrated crystals to the anhydrous brown form. After all of the crystals appear to be brown, continue heating gently and, judiciously, for an additional two minutes. Gently remove the evaporating dish from the hot plate all allow it to cool for about 10 minutes. Slowly roll the brown crystals around the evaporating dish. If some green crystals remain, repeat the heating process. Finally, weigh the cool evaporating dish and its contents accurately.

Transfer the brown crystals in the evaporating dish to an empty 50-mL beaker. Rinse out the evaporating dish with two 5- to 7-mL portions of distilled water, and add the rinsings to the beaker. Swirl the beaker gently to dissolve the brown solid. The color will change to green as the copper ions are rehydrated. Measure out about 20 cm of 20-gauge aluminum wire (~0.25 g) and form the wire into a loose spiral coil. Put the coil into the solution so that it is completely immersed. Within a few moments you will observe some evolution of H_2, hydrogen gas, and the formation of copper metal on the Al wire. As the copper ions are reduced, the color of the solution will fade. The Al metal wire will be slowly oxidized and enter the solution as aluminum ions. (The hydrogen gas is formed as the aluminum reduces water in the slightly acidic copper solution.)

When the reaction is complete, which will take about 30 minutes, the solution will be colorless, and most of the copper metal that was produced will be on the Al wire. Add 10–15 drops of 6 M HCl to dissolve any insoluble aluminum salts and clear up the solution. Please note that HCl is corrosive and must be handled with care. Use your glass stirring rod to remove the copper from the wire as completely as you can. Slide the unreacted aluminum wire up the wall of the beaker with your stirring rod, and, while the wire is hanging from the rod, rinse off any remaining Cu particles with water from your wash bottle. If necessary, complete the removal of the Cu with a drop or two of 6 M HCl added directly to the wire. Put the wire aside; it has done its duty.

In the beaker you now have the metallic copper produced in the reaction, in a solution containing an aluminum salt. In the next step we will use a Buchner funnel to separate the copper from the solution. Weigh accurately a dry piece of filter paper that will fit in the Buchner funnel, and record its mass. Put the paper in the funnel, and apply light suction as you add a few mL of water to ensure a good seal. With suction on, decant the solution into the funnel. Wash the copper metal thoroughly with distilled water, breaking up any copper particles with your stirring rod. Transfer the wash and the copper to the filter funnel. Wash any remaining copper into the funnel with water from your wash bottle. **All** of the copper must be transferred to the funnel. Rinse the copper on the paper once again with water. Turn off the suction. Add 10 mL of 95% ethanol to the funnel, and after a minute or so turn on the suction. Draw air through the funnel for about 5 minutes. With your spatula, lift the edge of the paper, and carefully lift the paper and the copper from the funnel. Dry the paper with copper on an watch glass for at least 5 minutes. Allow it to cool to room temperature and then weigh it accurately.

DISPOSAL OF REACTION PRODUCTS. Dispose of the liquid waste and copper produced in the experiment as directed by your instructor.

Determination of a Chemical Formula

name _____ date _____ instructor _____

section _____ partner's name _____

Write your answers in pen and include units.

Pre-Laboratory Questions

1. Iron(II) chloride tetrahydrate, $FeCl_2.4H_2O$ has a formula weight of 198.1 g/mol.
   (a) If we removed all the water from 198.1 g of $FeCl_2.4H_2O$, how much water would we get?

   (b) After that, if we removed all the chlorine, how much iron would remain?

2. Define molar mass.

3. What is the color of copper chloride hydrate?

4. What are the color changes you can see when you heat the sample? Why?

5. What is an *anhydrous* compound?

6. A compound is found to contain 24.8% carbon, 2.0% hydrogen and 73.2% chlorine with a molecular mass of 96.9 g/mol. What is the molecular formula?

_____ _____ _____
name   date   instructor

_____ _____
section   partner's name

Write your answers in pen and include units.

Report Form

Determination of a Chemical Formula

Atomic Masses: Cu_____   Cl_____   H_____   O_____

DATA

Mass of evaporating dish _____ g

Mass of evaporating dish and hydrated sample _____ g

Mass of hydrated sample _____ g

Mass of evaporating dish and dehydrated sample _____ g

Mass of dehydrated sample _____ g

Mass of filter paper _____ g

Mass of filter paper and copper _____ g

Mass of copper _____ g

    No. moles of copper *(show calculation)* _____ mol

Mass of water evolved _____ g

    No. moles of water *(show calculation)* _____ mol

Mass of chlorine in sample _____ g

    No. moles of chlorine *(show calculation)* _____ mol

DETERMINE THE FOLLOWING BASED ON MEASUREMENTS RECORDED

Mole Ratio (show calculations)

    Chlorine : copper in sample   _____

    Water : copper in hydrated sample   _____

Formula of dehydrated sample (round to nearest interger)   _____

Formula of hydrated sample   _____

name _____ date _____ instructor _____

section _____ partner's Name _____

Write your answers in pen and include units.

Post Lab Questions: Determination of a Chemical Formula

1. To find the mass of a mole of an element, one looks up the atomic mass of the element in a table of atomic masses (the Periodic Table). The molar mass of an element is simply the mass in grams of that element that is numerically equal to its atomic mass. For a compound substance, the molar mass is equal to the mass in grams that is numerically equal to the sum of the atomic masses in the formula of the substance. Find the molar mass of

   Cu _____ g   Cl _____ g   H _____ g   O _____ g H_2O _____ g

2. If one can find the ratio of the number of moles of the elements in a compound to one another, one can find the formula of the compound. In a certain compound of copper and oxygen, Cu_xO_y, we find that a sample weighing 0.6349 g contains 0.5639 g Cu.

   a. How many moles of Cu are there in the sample?

   $$\left(\text{No. moles} = \frac{\text{mass Cu}}{\text{molar mass Cu}}\right)$$

   _____ moles

   b. How many grams of O are there in the sample? (The mass of the sample equals the mass of Cu plus the mass of O.)

   _____ g

   c. How many moles of O are there in the sample?

   _____ moles

   d. What is the mole ratio (no. moles Cu/no. moles O) in the sample?

   _____ : 1

   e. What is the formula of the oxide? (The atom ratio equals the mole ratio, and is expressed using the smallest integers possible.)

   _____

   f. What is the molar mass of the copper oxide?

   _____ g/mol

69

Lab 6

Solubility

Prepared by Lee R. Summerlin, University of Alabama

When we use chemicals in the laboratory, it is usually more convenient to use them as solutions. *A solution is a homogeneous mixture of a solvent in which one or more of another component, the solute, is dissolved.* For example, in a solution of sodium chloride, solid sodium chloride (the solute) is dissolved in water, the solvent. You cannot make a solution by just dissolving any chemical in any solvent. Some chemicals will not dissolve in water—we say they are *insoluble*. Some chemicals will dissolve only in certain solvents, such as acetone, toluene or benzene. The chemical nature of the solute and solvent determine whether or not one will dissolve in the other.

In this experiment, you will explore the ability of some solutes and solvents to form solutions. You will also explore some factors that effect the solubility of chemicals, such as temperature and the chemical nature of the solute and solvent.

First, however, you will carry out some chemical reactions that result in the production of chemical compounds that are insoluble in water. These are called *precipitates*. Using a solubility chart, you will be able to predict whether a compound you produce is soluble or not, and learn to write a balanced chemical equation to represent the reaction that produces an insoluble compound.

With a little practice, you can make predictions like any other chemist about the solubility of ionic compounds.

Compounds containing	are	except
Alkali metals (Li^+, Na^+, K^+, Rb^+, Cs^+)	soluble	
Ammonium (NH_4^+)	soluble	
Nitrate (NO_3^-)	soluble	
Chlorate (ClO_3^-)	soluble	
Perchlorate (ClO_4^-)	soluble	
Acetate ($C_2H_3O_2^-$)	soluble	
Chlorides (Cl^-), Bromides (Br^-), Iodides (I^-)	soluble	Lead (Pb^{2+}), Silver (Ag^+), and Mercury (Hg_2^{2+})
Sulfates (SO_4^{2-})	soluble	Strontium (Sr^{2+}), Barium (Ba^{2+}), Mercury (Hg_2^{2+}), Mercury (Hg^{2+}), and Lead (Pb^{2+})
Hydroxide (OH^-)	insoluble	Alkali metals, Ca^{2+}, Ba^{2+}, and Sr^{2+}
Sulfites (SO_3^{2-}), Chromates (CrO_4^{2-}), Carbonates (CO_3^{2-}), and Phosphates (PO_4^{2-})	insoluble	Alkali metals and NH_4^+
Sulfides (S^{2-})	insoluble	Alkali metals, Beryllium (Be^{2+}), Magnesium (Mg^{2+}), Calcium (Ca^{2+}), Strontium (Sr^{2+}), Barium (Ba^{2+}), and NH_4^+

A. NAME THAT PRECIPITATE

In this experiment, you will mix a series of solution-pairs. Using the solubility chart, you can predict if a precipitate will form, and you will be able to name the precipitate. All of the solutions that you will use are solutions of ionic compounds.

Procedure

1. Place 10 drops of saturated calcium chloride ($CaCl_2$) solution in a small test tube.

    What ions are present in this solution? _____ and _____.

2. Add 5 drops of saturated potassium carbonate, K_2CO_3, solution to the tube.

    What ions are present in this solution? _____ and _____.

    What do you observe? _____

    This is the reaction, showing all ions in the reactants and all possible products:

    $$Ca^{2+} + 2\ Cl^- + 2\ K^+ + CO_3^{2-} \rightarrow 2\ KCl + CaCO_3$$

    Which is the white precipitate you observed in this experiment? According to the "Rules" of solubility, all alkali metal compounds are soluble. Thus, the white precipitate could not be KCl. Also, according to the chart, carbonates combined with calcium, $CaCO_3$, are insoluble. Therefore, this must be the white precipitate!

    We can now rewrite the equation, showing *only* those ions that take part in the formation of calcium carbonate, and eliminating the non-participating, or "spectator" ions from the reaction:

    $$Ca^{2+}_{(aq)} + CO_3^{2-}_{(aq)} \rightarrow CaCO_{3(s)}$$

3. Repeat the procedure for each of the pairs of ionic solutions shown below. For each mixture of the two solutions, record your observations and answers on the Report Form using the terms 'soluble', 'insoluble', 'cloudy', 'effervescent', and 'precipitate formation',

   a. Barium chloride ($BaCl_2$) and sodium carbonate (Na_2CO_3)
   b. Nickel chloride ($NiCl_2$) and sodium carbonate (Na_2CO_3)
   c. Sodium carbonate (Na_2CO_3) and potassium iodide (KI)
   d. Barium chloride ($BaCl_2$) and sodium sulfate (Na_2SO_4)
   e. Copper (II) chloride ($CuCl_2$) and sodium carbonate (Na_2CO_3)

   POUR THE CONTENTS OF THESE TUBES INTO THE SPECIAL WASTE JAR LOCATED IN THE HOOD. RINSE YOUR TUBES AND POUR THE RINSE WATER INTO THE WASTE JAR ALSO.

B. FACTORS CAUSING SOLUBILITY

1. Temperature

   The solubility of most substances increases when the temperature is increased, although the increase in solubility will vary according to the nature of the solute. It is common experience that more sugar, for example, dissolves in hot water than in cold water.

   You will study the effect temperature has on the solubility of one ionic compounds.

Procedure

   a. Place 5 mL of water in each of two test tubes. Label both tubes "NaCl".
   b. Accurately weigh 0.5 g samples of solid NaCl.
   c. Place each weighed sample into a test tube and thump the tubes lightly for a few seconds. Note the appearance of the mixtures and record this on your Report Form.
   d. Place one test tubes in a beaker of boiling water for 15 minutes and place the other test tube in a beaker with ice water. Suspend each test tube by clamping it to a ring stand. Record your observations on the Report Form.

2. "Like Dissolves Like:" Polarity and Solubility

   Some molecules, such as water, have an unequal distribution of positive and negative charge within their molecule, even though the molecule is overall neutral. We say that these molecules are *polar*. Most ionic compounds are polar. Some compounds, however, have an equal charge distribution, and are said to be *non-polar*. These include organic solvents such as benzene and carbon tetrachloride. The iodine molecule is a non-polar molecule. It has been observed that non-polar compounds are more likely to be soluble in non-polar solvents, and that polar compounds are soluble in polar solvents. You will perform an experiment to test this generalization.

Procedure

a. Place 5 mL of water into each of 5 test tubes. Label the test tubes as follows:

   1. Iodine
   2. Sodium chloride
   3. Sugar
   4. Wax
   5. Cooking oil

b. Place a small amount of each substance (match-head size of the solids, 2-3 drops of the cooking oil) into its respective tube.
c. Mix the contents by thumping the tube with your fingers. Record your observations on the Report Form.
d. Repeat the experiment, but this time put 5 mL of toluene in each test tube instead of water. Add the substances listed above to their respective tubes.
e. Mix the tubes as you did before. Record your observations on the Report Form.

Dispose of your materials by rinsing them into the waste jars located in the hood.

WHAT DISSOLVES AND WHAT DOES NOT: SOME GENERAL RULES

After making a large number of observations on the solubility of compounds, chemists have come up with some helpful "rules" to determine if an ionic compound is soluble or insoluble in water. These generalizations are listed in the chart displayed earlier.

The chart is easy to read. For example, it tells us that compounds containing ammonium (NH_4^+) are soluble (no exceptions). This means that NH_4Cl, $(NH_4)_2S$, $(NH_4)_3PO_4$, and so on are all soluble. It also tells us that compounds containing hydroxide (OH^-) are insoluble, except the alkali metals, calcium, barium, and strontium. For example, magnesium hydroxide ($Mg(OH)_2$) is insoluble, but calcium hydroxide ($Ca(OH)_2$) is soluble.

name _____ date _____ instructor _____

section _____ partner's Name _____

Write your answers in pen and include units.

Pre-Laboratory Questions:

1. Imagine that saturated solutions of barium chloride ($BaCl_2$) and potassium sulfate (K_2SO_4) are mixed. A white precipitate forms, and the solution is found to contain potassium chloride.

    (a) What is the name of the precipitate?

    (b) Write an equation for the reaction that occurred.

    (c) Which are the spectator ions in this reaction?

2. Consider the compound naphthalene.

    Based in the rule "Like Dissolves Like", which should be a good solvent for naphthalene and why: benzene, water, acetone? *(Hint: Use your textbook and/or other resources to help you identify the structure of the named compounds)*

3. When mixing the ionic solution pairs, is the quantity of each solution added to the test tube important? Why or not?

4. Predict the water solubility of following compounds: RbCl, NH_4NO_3, PbS, $BaSO_4$, AgBr

5. Define a *homogenous solution*

name _____ date _____ instructor _____

section _____ partner's name _____

Write your answers in pen and include units.

Solubility

Report Form

A. Name That Precipitate

Reaction pair	Ions in solution	Name of precipitate	Net ionic reaction	Observations
$CaCl_2$ and K_2CO_3				
$BaCl_2$ and Na_2CO_3				
$NiCl_2$ and Na_2CO_3				
Na_2CO_3 and KI				
$BaCl_2$ and Na_2SO_4				
$CuCl_2$ and Na_2CO_3				

B. Factors Effecting Solubility

1. Temperature

Ionic compound	Describe mixture at room temperature	Describe mixture after heating	Describe mixture after cooling
NaCl			

© 2006 Cengage Learning

77

2. "Like Dissolves Like:" Polarity and Solubility

   A. Polar Solvent: Water

Solute	Observations
Iodine	
Sodium chloride	
Sugar	
Wax	
Cooking oil	

1. What patterns do you observe regarding the polar and non-polar characteristics of these solutes?

2. What effect do you think heating would have on the solubilities of these substances?

   B. Non-Polar Solvent: Toluene

Solute	Observations
Iodine	
Sodium chloride	
Sugar	
Wax	
Cooking oil	

Based on the 'likes dissolve likes' concept, what generalizations can you draw regarding the solubilities of the following substances:

1. Which of the solutes are polar?

2. Which are non-polar?

Post-Lab Questions

_____   _____ _____
name   date   instructor

_____   _____
section   partner's name

Write your answers in pen and include units.

1. Draw a structure for water, showing that it is a polar solvent.

2. Draw a structure for CCl_4, carbon tetrachloride. Is this a polar or non-polar solvent?

3. Generally, what effect does increasing temperature have on the solubility of a substance?

4. Why are "spectator ions" not included in the net ionic reaction?

5. What did you observe in this experiment that tells us whether wax is polar or non-polar?

6. Why does a precipitate form when certain solutions are mixed?

7. We take special precautions to discard the precipitates formed in this experiment. Why?

8. Give evidence from your everyday experience that increasing the temperature increases the solubility of some substances.

9. What evidence from your everyday experience can you give that "like dissolves like?"

Lab 7

Types of Chemical Reactions

Performance Goals

1 Carry out various chemical reactions.
2 Demonstrate that during chemical reactions mass is conserved.

CHEMICAL OVERVIEW

Chemical reactions can be classified as:

   a) Combination or synthesis reactions in which two or more substances combine to form a single product.
   b) Decomposition reactions, which are the opposite of the combination reactions, in that a compound breaks down into simpler substances.
   c) Complete oxidation (burning) of organic compounds. In these reactions an organic compound reacts with oxygen yielding carbon dioxide, $CO_2(g)$ and water, $H_2O(g)$ or $H_2O(l)$.
   d) Precipitation reactions, when the cation from one compound reacts with the anion of another compound yielding a solid product (precipitate). These reactions are also called double replacement or ion combination reactions since ions of the two reactants appear to change partners.
   e) Oxidation–reduction reactions, during which one of the reactants gives off electrons (gets oxidized) and the other gains electrons (gets reduced).
   f) Acid–base reactions, also called neutralization reactions, in which an acid reacts with a base yielding a salt and (usually) water.

In this experiment, we will carry out several different reactions, starting and ending with metallic copper. These reactions can be summarized as follows:

$$Cu \xrightarrow{HNO_3} Cu(NO_3)_2 \xrightarrow{NaOH} Cu(OH)_2 \xrightarrow{heat} CuO \xrightarrow{HCl} CuCl_2 \xrightarrow{Zn} Cu$$

SAFETY PRECAUTIONS

In this experiment you will use fairly concentrated acids and bases. When in contact with skin, most of these chemicals cause severe burns if not removed promptly. Always wear goggles when working with these chemicals. Reacting metal with nitric acid should *only* be carried out *in the hood*. Be careful when using a boiling water bath. Replenish the water from time to time as it becomes necessary.

PROCEDURE

1. DISSOLUTION OF COPPER

   A. Weigh a clean and dry 25-mL Erlenmeyer flask on a milligram balance. Record this value on the work page.

B. Place about 100 mg of metallic copper (wire or granules) into the flask. Weigh the metal and flask to the milligram and record the mass.

C. *In the hood,* add 2 mL of 6 M nitric acid, HNO_3, to the flask and warm the contents on a hot plate. Brown vapors will form as the metal dissolves. Continue heating until no more brown fumes exist over the solution. Be sure not to evaporate all of the liquid. If needed, add two more milliliters of HNO_3. Allow the solution to cool to room temperature, then add 2 mL of deionized water.

2. PREPARATION OF COPPER(II) HYDROXIDE

To the solution prepared above, carefully add 6 M sodium hydroxide, NaOH, drop by drop, until the solution is basic to litmus (red paper turns blue). You can use magnetic stirring or swirl the contents of the flask while adding the NaOH. Do not dip the litmus paper directly into the solution. Instead, stir the solution with a glass rod and then touch the wet end of the rod to the paper. You should see a blue spot on the red paper when the solution is basic.

3. PREPARATION OF COPPER(II) OXIDE

A. While stirring, heat the flask and its contents in a boiling water bath or on a magnetic stirring hot plate. In about 5 minutes, the blue $Cu(OH)_2$ will be converted to the black copper(II) oxide. If this does not occur, check your solution; it may not be basic enough. Swirl your solution and add more NaOH, then check with the litmus paper.

B. Allow the mixture to cool to room temperature. Remove the magnetic stirrer, if used. Rinse with a small amount of deionized water, collecting the rinse in the Erlenmeyer flask.

C. Set up a vacuum filtration apparatus using a small Büchner funnel. Place a small filter paper into the funnel, moisten it with a small amount of deionized water, and start the vacuum. This will "seat" the filter paper and eliminates leakage around its edges. After this point, *do not* shut off the vacuum until the filtration is finished.

D. Transfer the black precipitate into the funnel, rinse the flask with 1 to 2 mL of deionized water, and pour into the funnel. The filtration may be a bit slow toward the end, due to small particles plugging up the filter paper pores. Wash the precipitate with 1 to 2 mL of deionized water. Discard the filtrate.

4. CONVERTING COPPER(II) OXIDE TO COPPER(II) CHLORIDE

Pour 6 mL of 6 M hydrochloric acid, HCl, into a 50-mL beaker. Using a spatula, transfer the black precipitate *and* the filter paper to the acid solution. Do not let the metal spatula come in contact with the acid. Stir the mixture with a glass stirring rod until the precipitate is completely dissolved. If needed, heat the solution on a hot plate. Remove the filter paper, using the glass rod, and rinse it with 1 to 2 mL of deionized water, adding the rinse to the green solution. *Do not* use metal forceps or tweezers because they will contaminate the solution. If some precipitate is stuck on the funnel, hold it over the beaker and rinse it with 1 to 2 mL of 6 M hydrochloric acid solution. Rinse the funnel with 1 to 2 mL of deionized water. This rinse should also be collected in the beaker.

5. RECOVERING THE METALLIC COPPER

A. Weigh about 200 mg of zinc powder on a piece of preweighed weighing paper.

B. *In the hood,* **very** slowly, add a small amount of zinc powder to the copper(II) chloride solution. Stir after each addition. You will observe the formation of copper metal particles and vigorous evolution of hydrogen gas. This step is very critical, because too rapid formation of copper globules tends to enclose some of the unreacted zinc powder. This will result in an unrealistically high yield for the experiment.

C. Test for completeness of the reaction by adding 2 to 3 drops of your solution to 10 drops of concentrated ammonia, NH_3, in a small test tube. If a blue color appears, the reaction is not yet complete. Add a few more *small* portions of zinc powder and test again. Another indication that all the copper has been removed is the fact that the green solution turns colorless.

D. After the reaction is complete, add 5 mL of 3 M hydrochloric acid to the solution in the beaker and stir with a glass rod. This will hasten the removal of excess zinc present in your mixture. Metallic copper does not react with hydrochloric acid. Allow the solution to stand for 5 minutes, stirring occasionally.

E. Place a small funnel into a 250-mL Erlenmeyer flask or secure it with a clamp over a beaker. Weigh a piece of filter paper and place it in the funnel. First, pour the solution into the funnel, then transfer the solid copper. Use deionized water to rinse the beaker and be sure that all solid has been collected in the funnel. Wash the copper twice with 2-mL portions of deionized water.

F. Remove the filter paper and copper from the funnel, spread it out on a watch glass, and allow it to air dry. At the beginning of the next laboratory period weigh the copper and filter paper to the milligram and record the mass on the work page. Calculate the percentage of recovery (yield).

Types of Chemical Reactions

Pre-Laboratory Questions:

_____   _____ _____
name   date   instructor

_____   _____
section   partner's name

Write your answers in pen and include units.

Classify the following reactions as either: *combination, decomposition, combustion, precipitation (metathesis) or redox reactions*:

1. $C\,(s) + 4\,HNO_3\,(aq) \rightarrow 4\,NO_2\,(g) + 2\,H_2O\,(l) + CO_2\,(g)$

2. $HCl\,(aq) + NH_3\,(aq) \rightarrow NH_4Cl\,(aq)$

3. $2\,HI\,(g) \rightarrow H_2\,(g) + I_2\,(g)$

4. $Cu(NO_3)_2\,(aq) + Na_2S\,(aq) \rightarrow CuS\,(s) + 2\,NaNO_3\,(aq)$

5. $Zn\,(s) + 2\,AgNO_3\,(aq) \rightarrow Zn(NO_3)_2\,(aq) + 2\,Ag\,(s)$

6. $H_2SO_3\,(aq) + 2\,KOH\,(aq) \rightarrow K_2SO_3\,(aq) + 2\,H_2O\,(l)$

Report Form

_____   _____ _____
name   *date*   *instructor*

_____   _____
section   *partner's name*

Write your answers in pen and include units.

MASS DATA

Mass of flask (g)	
Mass of flask + Cu (g)	
Mass of weighing paper (g)	
Mass of weighing paper + Zn (g)	
Mass of filter paper (g)	
Mass of filter paper + Cu (g)	

RESULTS

Mass of Cu, initial (g)	
Mass of Zn (g)	
Mass of Cu, recovered (g)	
Percent recovery	

Show all your calculations below:

Classify each reaction as a double replacement, synthesis, decomposition, precipitation, neutralization, or oxidation–reduction reaction:

Part 1. _____

Part 2. _____

Part 3. _____

Part 4. _____

Part 5. _____

Post Laboratory Questions

_____   _____   _____
name   date   instructor

_____   _____
section   partner's Name

Write your answers in pen and include units.

1. Balance each of the following equations and classify the reactions:

   a) $Cu(s) + HNO_3(aq) \rightarrow Cu(NO_3)_2(aq) + H_2O + NO_2(g)$

   _____

   b) $Cu(NO_3)_2(aq) + NaOH(aq) \rightarrow Cu(OH)_2(s) + NaNO_3(aq)$

   _____

   c) $Cu(OH)_2(s) \rightarrow CuO(s) + H_2O$

   _____

   d) $CuO(s) + HCl(aq) \rightarrow CuCl_2(aq) + H_2O$

   _____

   e) $CuCl_2(aq) + Zn(s) \rightarrow Cu(s) + ZnCl_2(aq)$

   _____

2. Define a precipitation reaction.

3. If you started with 0.108 g of copper and at the end of the experiment you had recovered 0.099 g, calculate the percent recovery.

Lab 8
Limestone Caves
Prepared by Dr. Ekkehard Sinn, Hazim Al-Zubaidi, and Brianna Galli

Introduction

Why is Limestone important?

Limestone is a common sedimentary rock found on all of the continents. Usually it is comprised of the minerals calcite and aragonite, which have the chemical formula $CaCO_3$ (calcium carbonate). Some limestone has magnesium, Mg, in place of some of the calcium. This is known as dolomite, which has the chemical formula $CaMg(CO_3)_2$.

Limestone is the main stone quarried in the United States, at over 2 billion tons per year - about 42% of all stone produced, far more than coal. Wisconsin and Indiana produce most of it (87%) but all of the 50 states are in the limestone business. Michigan has the largest limestone quarry; Missouri sits mostly on limestone and just a small fraction of this state is limestone-free and volcanic. Most of England, part of Germany, and the sea between the two sit on a single large slab of limestone.

Use – It is cut for use directly as building material, as a component in Portland cement, but its main use is as aggregate such as gravel to build roads. It has many other uses, e.g. in medicines, and also in whitewash. Producing quicklime, CaO, requires extreme heat, which is why it is also called "burnt lime" and releases carbon dioxide.

$$CaCO_3 \rightarrow CaO + CO_2 \ldots\ldots [1]$$

Some quicklime is sold, but most is used to make lime, $Ca(OH)_2$. The quicklime reacts vigorously when it is "slaked" (hydrated). This is used in agriculture and can also be used to make mixed calcium hydroxide/carbonate which is called whitewash. Whitewash soaks into wood and is a cheap way to preserve wooden buildings and make them look prettier; CO_2 from the air turns back into a hard $CaCO_3$ coating. Since it takes CO_2 from the air, it is an environmentally friendly "green" paint, made famous in Mark Twain's "Tom Sawyer", when Tom has the chore of whitewashing a fence and persuades his friends to pay him for the privilege of sharing the work.

Whitewashing the Old House, painting by L.A. Ring 1908 from Statens Museum for Kunst, Copenhagen. Public domain permission

Where does Limestone come from?

Over millions of years, small marine animals form hard shells using dissolved $Ca^{2+}_{(aq)}$ and $HCO_3^-{}_{(aq)}$ to form insoluble $CaCO_3$. $Ca(HCO_3)_2$ is soluble in water, *but it can only exist in solution*. If it is warmed or if the solution dries, it quickly loses CO_2 and water as depicted in the chemical equation below:

$$Ca(HCO_3)_{2\ (aq)} \rightarrow CaCO_{3(s)} + H_2O_{(g\ or\ l)} + CO_{2(g)}\ldots\ [2]$$

$CaCO_3$ is a sedimentary rock of biological origin. When the organisms die, the organic matter gradually decays and the fragments of their calcium carbonate shells aggregate into limestone.

What about Limestone caves?

Limestone is insoluble in neutral water, but dissolves in acidic water. The process in equation [2] above can reverse with natural acid rain: $H_2O + CO_2 \rightleftharpoons H^+ + HCO_3^-{}_{(aq)}$ (weak acid). So, for example, Meramec Carverns in Missouri existed for about 400 million years, gradually forming from water containing a little CO_2 which dissolves out limestone and takes it away as $Ca(HCO_3)_2$ solution. The caverns were used in the US civil war and as Jesse James' hideout and then as a tourist attraction. Over 5 miles of cave passage are mapped and all of it accessible.

Most limestone caves are completely dark and have their own micro-ecology of animals that do not exist anywhere else, e.g. sightless fish, insects and amphibians. They also represent a historical record of above-ground animal populations as skeletons of individuals that accidentally entered but never got out again. Marvel Cave in Missouri was discovered by the Osage Indians when one tribesman fell in through a narrow sinkhole. It is now a tourist attraction. Some limestone caves discovered long ago were treated rather destructively. Later they came to be seen as natural wonders of the world, for scientific study and recreation.

In recent times (the "Anthropocene") acid rain has become more acidic, and dissolves limestone much more readily. This not only accelerates the destruction of limestone caves but can play havoc with limestone in commercial applications, e.g. sink-holes under streets and buildings.

We will investigate how insoluble limestone is by making it from $CaCl_2$ and Na_2CO_3 solutions. We will also form $Ca(HCO_3)_2$ solutions from $CaCl_2$ and Na_2CO_3 and see how fragile the calcium bicarbonate solution is – it will probably precipitate out by itself but if it does not, we will warm it slightly. We will check what happens when we expose limestone to stronger acid, e.g. HNO_3, as found in acid rain. The rate of these geochemical reactions varies but it is very slow, and we do not have hundreds or thousands of years available to finish the lab. Therefore we will use more concentrated acid than naturally found in acid rain, to simulate acid rain working on a highly accelerated timescale.

Useful sources: US National Park Service, results of chemical, geological, and biological study;

US Geological Survey, many resources including maps;
Meramec Cave History, discovery, use and policy change;

PROCEDURE

You will be making solutions of sodium carbonate, sodium bicarbonate, calcium chloride, and calcium carbonate. You will then perform solubility tests to simulate the formation of solution caves. From your results, you will connect a relationship of physical property to a geochemical phenomenon.

Part 1 Calcium carbonate precipitation

A. Preparation
To prepare 50 mL solutions of Na_2CO_3, $NaHCO_3$, and $CaCl_2$, you will need to recall dimensional analysis. First, identify what the goal is, what information you have, and what you need to find out. Collaborate to outline a procedure. Once approved, prepare your solutions.

B. Reaction
You will mix calcium chloride separately with sodium carbonate and sodium bicarbonate. You need to write the chemical equations to predict the products. Collaborate to outline a procedure. Once approved, conduct the reactions.

Part 2 Calcium carbonate conversion

Prepare 100 mL of 0.01 M calcium carbonate and manipulate the solution to improve dissolution.
1. Obtain a 250 mL Erlenmeyer flask.
2. Weigh the flask using an analytical balance.
3. Obtain solid calcium carbonate and weigh out 0.1 g into the flask with a scoopula.
4. Reweigh the flask to determine mass of the solid.
5. Obtain a 100 mL graduated cylinder and fill with DI water.
6. Add water in specific increments to the flask. Mix the solution and record observations
7. After 5 minutes: if available and with supervision, add a small piece of dry ice to the solution and record observations. If still cloudy, add a larger piece of dry ice.
8. After 5 minutes: add acid incrementally to the solution. Mix the solution and record observations

If time permits, repeat this procedure but allow a longer duration in part 7 and with more vigorous stirring.

> Safety Notice
> *Dry ice* is solid carbon dioxide, which sublimes at temperatures greater than -78.5 °C. Be sure to take proper thermal precautions by wearing gloves and avoiding bodily contact. Use wooden or plastic tools to handle pieces of dry ice.
> *Hydrochloric acid* is corrosive and can cause skin and eye damage. Goggles and gloves must be worn when HCl is present.

Calculation Practice

Calculate the molecular weight for a compound from its molecular formula:

Example:

    Calcium bicarbonate

    $Ca(HCO_3)_2$ 1 Ca + 2 H + 2 C + 6 O

                             = 1*(40.078) + 2*(1.0079) + 2*(12.011) + 6*(15.9994)

                             = 162.112 g/mol

                             Note that this compound only exists in aqueous solution and not in solid form.

Compounds:

    sodium carbonate

    sodium bicarbonate

    calcium chloride

    calcium carbonate

CHEM 1110 Limestone Cave Prelaboratory Exercise

Name _____ Date _____

Section _____ Instructor _____

Write your answers in pen.

1. What is the chemical name for
    (a)   Limestone:

    (b)   Quicklime:

    (c)   Lime:

2. Write a balanced equation for making "slaked lime" from "quicklime".

3. Complete the following reactions:
    (a) $CaCl_2 + Na_2CO_3 \longrightarrow$

    (b) $CaCl_2 + NaHCO_3 \longrightarrow$

    (c) $CaCO_3 + CO_2 + H_2O \longrightarrow$

4. Why can magnesium, Mg, replace calcium, Ca, in limestone to form dolomite (see periodic table)?

5. Describe the difference between compounds that are anhydrous versus those that have waters of hydration.

6. Reference your course text or the internet to fill out the bold sections of the solubility chart.

Group	Compounds containing	Exceptions
Soluble Salts		
Group 1A cations, alkali metals	Li^+, Na^+, K^+, Rb^+, Cs^+	None
Ammonium salts	NH_4^+	None
Nitrate, chlorate, perchlorate, and acetate salts		None
Chloride, bromide, and iodide salts		Lead (Pb^{2+}), Silver (Ag^+), Mercury (Hg_2^{2+})
Sulfates	SO_4^{2-}	Barium (Ba^{2+}), Mercury (Hg_2^{2+} and Hg^{2+}), Lead (Pb^{2+}), Strontium (Sr^{2+})
Insoluble Salts		
Hydroxides	OH^-	
Sulfites, chromates, carbonates, and phosphates		
Sulfides	S^{2-}	Alkali metals/group 1A cations, NH_4^+, and group 2A metals

7. Look up chemical safety information for the chemicals used in this lab. Reference the safety data sheets (SDS) and list the hazards associated with each chemical.

    sodium carbonate

    sodium bicarbonate

    calcium chloride

    calcium carbonate

8. Visit the website of a cave system in the United States such as Wind Cave National Park, Jewel Cave National Park, Mammoth Cave National Park, or Karchner Caverns State Park. Write 3 things you found interesting about the cave or the park location.

    Cave system:
    Website:
    Interesting information:
        1)
        2)
        3)

CHEM 1110 Limestone Cave Experimental Design

Name _____ Date _____
Section _____ Instructor _____
Day/Time _____ Lab Partner _____

Write your answers in pen.

Part 1-A Calcium carbonate precipitation: Preparation of 50 mL solutions

Goal (rephrase and expand the Part 1-A title above; see example in Part 1-B):

Information you have:

Information to collect:

Sample calculations (show units!):
Confirm your calculations with a lab partner and with your instructor. Record your final answers in Table 1 of the data sheet.

Decide on a method to prepare these solutions and outline a procedure:

1.
2.
3.
4.
5.
6.
7.
8.
9.
10.

Once approved, prepare the solutions and continue to Part 1-B. If you need remake these solutions, you can record the mass of solid chemical in Table 2. TA initial _____

Part 1-B Calcium carbonate precipitation: Reaction

Goal: To understand the separate chemical reactions of $CaCl_2$ with Na_2CO_3 and $NaHCO_3$, predict the products, carry out the reaction and observe physical changes.

Information you have:

Information to collect:

Sample work (write out 2 chemical equations):

Confirm your chemical reaction equations with a lab partner and with your instructor.

1. What types of reactions are these?

2. What physical changes do you expect to occur?

Decide on a method to react calcium chloride separately with sodium carbonate and sodium bicarbonate. Confirm with a lab partner and your instructor before conducting. Record your observations in Table 3 in the data sheet.

Outline your procedure below:

1.
2.
3.
4.
5.
6.
7.
8.
9.
10.

Part 2 Calcium carbonate reversible conversion

Goal (rephrase and expand the Part 2 title above; see example in Part 1-B):

Information you have:

Information to collect:

Sample work (chemical equation of $CaCO_3 + H_2O + CO_2 \rightarrow ...$):

Confirm your chemical reaction equations with a lab partner and with your instructor.

What is the concentration of the calcium carbonate solution you prepared? Show calculation.

Follow the procedure provided and record your observations in Table 4 in the data sheet. Recall safety hazards and consult your instructor for any additional standard operating procedures, including proper disposal.

Data Sheet

Record the mass of solid chemical used to make the specified volume, 50 mL

Table 1: Solution Preparation

Compound	Concentration	Calculated Mass	Actual Mass (to 0.001 g)
Na_2CO_3 Sodium carbonate	0.20 M		
$NaHCO_3$ Sodium bicarbonate	0.20 M		
$CaCl_2$ Calcium chloride	0.10 M		

If you need to make different solutions, show your work here.

Table 2: Solution Preparation, revised

Compound	Concentration	Calculated Mass	Actual Mass (to 0.001 g)
Na_2CO_3 Sodium carbonate			
$NaHCO_3$ Sodium bicarbonate			
$CaCl_2$ Calcium chloride			

Table 3: Facilitate the reactions, then record observations for changes in appearance, precipitate formation, effervescence, solubility, pH, and rate of change.

Reaction	Initial Observations	Observations after 5 minutes
0.2 M Na_2CO_3 and 0.1 M $CaCl_2$		
0.2 M $NaHCO_3$ and 0.1 M $CaCl_2$		

Table 4: Prepare the calcium carbonate solution as instructed, then record observations for conversion to calcium bicarbonate	
100 mL of 0.01 M $CaCO_3$	Calculated mass:
	Actual mass (to 0.001 g):
Method	Observations
Add water to 100 mL solution in the 250 mL E. flask • 10 mL • 25 mL • 50 mL	
Add small piece of dry ice, carefully • _____ g	
Add second piece if necessary • _____ g	
Add HCl, carefully • 6 drops • _____ mL	

This page is intentionally left blank.

Post Laboratory Reflection

1. Write equation [2] in the net ionic form.

2. Describe what happens when strong acid acts on limestone.

    (a)   Write a molecular equation

    (b)   Write a net ionic equation

3. What are the consequences for marble statues in acid rain?

4. *Connecting the dots*: In a developed limestone cave, water continues to seep in from the ceiling, slowly forming stalactites. A "living" stalactite is wet, and has a pointed tip where water hesitates before falling. Meanwhile, stalagmites often grow upwards from the floor of the cave, usually right below a stalactite from water dripping off the stalactite.

Describe the chemical process in the formation of:
    a. Limestone caves

    b. Stalactites and stalagmites

    Why do the stalactites have pointed tips while the stalagmites usually do not?

5. Find the solubility in water (grams per Liter) at 20°C for each of the following:

   a. NaHCO$_3$

   b. Na$_2$CO$_3$

   c. CaCl$_2$

   d. Ca(HCO$_3$)$_2$

   e. CaCO$_3$

   Cite your source:

6. Find the commercial cost (dollar per kilogram) each of the following:
   (Hint: look at the websites of manufacturers like Sigma Aldrich for ACS grade).

   a. NaHCO$_3$

   b. Na$_2$CO$_3$

   c. CaCl$_2$

   d. Ca(HCO$_3$)$_2$

   e. CaCO$_3$

   Cite your source:

   *If you cannot find a commercial solid powder for any one of the chemicals above, can you explain why it's not available? Did you find any suspicious product offerings?

Lab 9

REAC 2154

Analysis of a Carbonated Beverage

A Guided-Inquiry Experiment

Prepared by Vickie M. Williamson and M. Larry Peck, Texas A&M University

INTRODUCTION

Many acidic compounds occur in nature. Some examples of naturally occurring acids are the fatty acids, amino acids, and citric acid in foods and the hydrochloric acid of the stomach. Nitrogen and sulfur containing oxyacids are found in acid rain, battery acid and various cleaning agents. In this experiment you will determine the citric acid content of a carbonated beverage.

OBJECTIVES

You are to investigate the relationships in the reaction of sodium hydroxide and citric acid. You are to perfect several basic laboratory skills frequently used in the laboratory process known as titration. You are to titrate standard, known, and unknown solutions. An indicator will be used to indicate when the endpoint of the titration is reached. Ultimately, you will use your titration data to calculate the concentration of citric acid in a carbonated beverage.

CONCEPTS

Monoprotic acids have only one acidic hydrogen atom per molecule. Diprotic acids have two and triprotic acids have three acidic hydrogen atoms per molecule. Hydrogens in a formula may or may not be acidic. Only those hydrogens that can be released as hydrogen ions in aqueous solutions are considered to be acidic. The neutralization reaction occurring between a monoprotic acid and sodium hydroxide is described by the following equation.

$$\text{HA} + \text{NaOH} \rightarrow \text{NaA} + \text{H}_2\text{O}$$
$$\text{monoprotic acid} \quad \text{base} \quad \text{salt} \quad \text{water}$$

The neutralization reaction occurring between a diprotic acid and sodium hydroxide is:

$$\text{H}_2\text{A} + 2\text{NaOH} \rightarrow \text{Na}_2\text{A} + 2\text{H}_2\text{O}$$
$$\text{diprotic acid} \quad \text{base} \quad \text{salt} \quad \text{water}$$

The reaction with a triprotic acid is similar but requires 3 moles of base and produces 3 moles of water instead of 2. When an alkaline solution is added to a solution of an acid, the OH^- from the added alkaline material and the H^+ ions from the acid combine to form undissociated water. After sufficient base has been added to have all the acidic protons neutralized (the endpoint of the titration), any additional base will rapidly increase the hydroxide ion concentration because H^+ ions are no longer present to combine with OH^-. Phenolphthalein reacts with the excess hydroxide ions and forms a bright red solution. Phenolphthalein, before its reaction with OH^-, is colorless. The appearance of a very pale pink color indicates that the neutralization of the acid is complete and that very small excess of base has converted some of the phenolphthalein into its alkaline or red form. This stage in a titration is referred to as the endpoint.

A measured volume of acidic solution of a known concentration and type can be titrated by using a buret to carefully add a solution of sodium hydroxide to it. The volume of sodium hydroxide solution needed to just reach the endpoint can be measured. The moles of sodium hydroxide consumed in the reaction can be calculated. For monoprotic acids, each mole of sodium hydroxide consumed in the neutralization reaction neutralizes one mole of the monoprotic acid. That is: at the endpoint the number of moles of base equals the number of moles of acid. Since molarity can be written as moles per liter, the following equation can be derived.

$$V_{NaOH}(\text{in liters}) \times M_{NaOH} = \text{moles}_{NaOH} = \text{moles}_{HA} = V_{HA}(\text{in liters}) \times M_{HA}$$

During the standardization of your sodium hydroxide solution, the moles of monoprotic acid used and volume of sodium hydroxide solution needed to reach the endpoint will be known. The above equation can be used to calculate the concentration of the sodium hydroxide solution.

In this experiment you will standardize a sodium hydroxide solution that you will then use to titrate the citric acid in a carbonated beverage. Sodium hydroxide is a rather reactive substance. Sodium hydroxide is one of many compounds that is hygroscopic (they absorb water from the air). The moist surface of the sodium hydroxide pellets will also react with carbon dioxide present in the air. For these reasons, sodium hydroxide is seldom considered to be pure. In this experiment you will prepare a sodium hydroxide solution by diluting a solution of doubtful concentration. The exact molarity of this solution will be determined by standardization with an acid that can easily be obtained in high purity, will not absorb water, and will not react with carbon dioxide or other constituents of the atmosphere. These are all desirable traits of a primary standard. Potassium hydrogen phthalate (KHP), the mono-potassium salt of the diprotic phthalic acid, is such a primary standard. The amount of KHP required for the standardization of a sodium hydroxide solution is carefully weighed on an analytical balance. The KHP sample is dissolved in hot water, phenolphthalein is added, and sodium hydroxide solution is added until the endpoint is reached. Since KHP is a monoprotic acid, at this point equal number of moles of acid and base have reacted.

potassium hydrogen phthalate (KHP) → potassium sodium phthalate (KNaP)

The representation above of KHP has a six-sided ring of single and double bonded carbon atoms representing the benzene-like portion of the compound. In this type of representation a carbon atom is located at each corner of the six-sided ring. A hydrogen atom is attached to each carbon atom that does not have more than two carbon atoms. The condensed formula of KHP is $C_8H_5O_4K$.

For the standardization of NaOH solution, the equation below can be used to calculate the molarity of a sodium hydroxide solution. In that equation M_{NaOH} is calculated when "m" grams of the potassium hydrogen phthalate (KHP), molar mass (MM_{KHP}), and "V_L" volume (in liters) of sodium hydroxide solution required are known.

$$m/MM_{KHP} = moles_{KHP} = moles_{NaOH} = V_L M_{NaOH}$$

Once standardized, the sodium hydroxide solution can then be used, if used within a short period of time, to titrate a different acid solution such as the carbonated beverage in this experiment.

To give a carbonated beverage a slight sour or tart taste, many manufacturers include an acid as one of the ingredients. The two acids most commonly used are citric acid ($HOC(CH_2CO_2H)_2CO_2H$, which has a condensed formula of $C_6H_8O_7$) and phosphoric acid (H_3PO_4) or a partially neutralized form of phosphoric acid such as NaH_2PO_4. Dark sodas such as the colas and root beer nearly always contain phosphoric acid since it is believed to enhance the taste of the caramel used to obtain the dark color. The less dark sodas, such as 7-Up™, usually have a more citrus, fruit-like taste due to the presence of citric acid as the acidic ingredient.

citric acid   or $C_6H_8O_7$

TECHNIQUES

You will prepare solutions, use the balances, transfer solids and liquids, control a chemical reaction, use the hot plate, record observations and make several calculations associated with an acid-base titration. The results of all calculations are to be expressed with significant figures only. These are skills and techniques that are considered to be essential for anyone who is preparing to enter any lab science career.

ACTIVITIES

You will prepare a sodium hydroxide solution of an approximate molarity, standardize the solution using potassium hydrogen phthalate, and use the standardized solution to determine, via titration, the number of acidic hydrogen atoms per molecule for citric acid and the concentration of citric acid in a carbonated beverage.

CAUTION

Wear approved eye protection at all times in the laboratory. Sodium hydroxide is a strong base; handle it carefully and avoid contact with your skin. If contact has occurred, wash with plenty of tap water.

PROCEDURES

Preparation of the Soda Sample

1-1. Place in a 250-mL beaker about 90 mL of a clear soda. Determine the exact volume of the sample with a 100-mL graduated cylinder to the nearest 0.1 mL. Record the volume and identity of your sample.

1-2. Your sample may be carbonated. That is, it may contain a considerable amount of dissolved CO_2. The dissolved CO_2 will interfere with the cirtic acid determination. Therefore, you need to remove most of the dissolved carbon dioxide before you can analyze for citric acid (Procedure 1-3). Before removing the CO_2 determine the initial acidity of the beverage.

Obtain a strip of pH Indicator Paper. To determine the pH of your sample, dip a stirring rod into the soda, then touch the stirring rod to the indicator strip. Take the strip to the color chart posted by your instructor and compare the colors. Record your findings.

1-3. Heat your sample on a hot plate to near boiling. Allow it to cool. Remeasure its volume. Add distilled water to bring its volume back to its initial volume (Procedure 1-1). Determine its pH now by using a second indicator strip.

Preparation of 150 mL of 0.04 *M* NaOH

2-1. With the help of your instructor, recalculate the volume of 2.5 *M* sodium hydroxide solution needed to make 200 mL of a 0.04 *M* solution (question #5 on the Prelab).

2-2. Take a clean, but not necessarily dry, 10-mL graduated cylinder to the reagent area and obtain the volume of ~2.5 M sodium hydroxide calculated in Procedure 2-1. Pour the ~2.5 M NaOH solution obtained into a 500-mL Erlenmeyer flask containing about 100 mL of distilled water. Use 5 mL of distilled water to rinse the graduated cylinder. Add this wash to the flask. Add additional distilled water to the 500-mL Erlenmeyer flask until a total volume of 200 mL is reached. Mix thoroughly by carefully swirling. Stopper the flask and label it "~0.04 M NaOH".

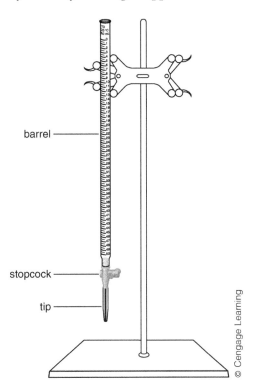

2-3. Set up a 50-mL buret. Check that it is clean, free of grease, and draining properly. Rinse the buret with three 5 mL portions of your ~0.04 M NaOH solution. Fill the buret with the NaOH solution to above the 0-mL mark. Drain some of the solution from the buret into a waste beaker to expel air from the buret tip. Allow the liquid level to drop below the zero mark. Restopper the flask containing the NaOH solution and set it out of the way.

Standardization of the 0.04 M NaOH Solution

3-1. In your Pre-Laboratory Exercises you calculated the mass of potassium hydrogen phthalate needed to neutralize 15 mL of an ~0.04 M NaOH solution. Have your instructor check your calculations.

3-2. Tare a clean, weighing boat on the analytical balance. Carefully add to the paper approximately the mass of potassium hydrogen phthalate (KHP) needed to neutralize 15 mL of a 0.04 M NaOH solution. The mass actually obtained may not be exactly the amount you needed. However, record the mass of the sample from the balance precisely. *Remember that the balance must be clean when you leave it.* Record the mass of your sample. Transfer your sample of KHP to a clean but not necessarily dry 250-mL Erlenmeyer flask.

3-3. Add approximately 100 mL distilled water to the flask. If the KHP doesn't dissolve with swirling, gently warm and carefully swirl the flask until all the phthalate has dissolved. Then remove the flask from the flame and allow it to partially cool.

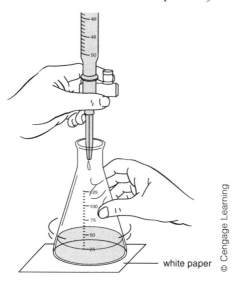

3-4. Add three or four drops of phenolphthalein indicator to the KHP solution in the flask and set the flask on a sheet of white paper under the buret. Read the volume in buret to the nearest 0.01 mL. Record the reading of the volume. While gently swirling the warm flask, slowly add the ~0.04 M NaOH solution to the KHP solution until a pale pink endpoint is observed and that persists for about a minute. Record the final buret reading. Empty the flask and rinse it with distilled water. Repeat the standardization.

Titration of Citric Acid

4-1. Refill the buret with the NaOH solution, if needed. Tare a clean piece of weighing boat on the analytical balance. Obtain 0.04 g of citric acid. Record the mass to the nearest milligram (0.001 g). Transfer the citric acid to a clean 250-mL Erlenmeyer flask.

4-2. Add approximately 100 mL distilled water to the flask. Carefully swirl and gently warm the flask until all of the solid has dissolved.

4-3. Add three or four drops of phenolphthalein indicator to the citric acid solution in the flask and set the flask on a sheet of white paper under the buret. Read the volume in the buret to the nearest 0.01 mL. Record the reading of the volume. While gently swirling the warm flask, slowly add the ~0.04 M NaOH solution until a pale pink endpoint is observed and persists for about a minute. Record the final buret reading. Empty the flask and rinse it with distilled water. Repeat the titration.

Titration of the Beverage

5-1. Make sure your 100-mL graduated cylinder is clean. Measure 35 mL of the degassed beverage solution to the nearest 0.1 mL. Pour into a clean, but not necessarily dry 250-mL beaker. Add four drops of phenolphthalein to the solution and titrate with the NaOH solution to a pale pink endpoint. Record the intial and final buret readings of the NaOH buret. Calculate the concentration of the citric acid in terms of both molarity and percent concentration by mass. If time allows, repeat the titration.

Clean-Up

6-1. All solutions may be safely flushed down the drain with large amounts of water.

6-2. All glassware must be rinsed and returned to their proper place.

6-3. Clean the burets with a buret brush and then rinse before putting away. Have your instructor sign your notebook and Report Form.

Name (Print) _____ Date (of Lab Meeting) _____ Instructor _____

Course/Section _____

Pre-Laboratory Exercises

1. Define:

concentration (no equations, please)

acid

endpoint

standardization of a solution

2. Calculate the molar mass of citric acid.

3. Sometimes there is an air bubble in the stopcock of the buret. Why do you want to remove the air bubble before starting a titration?

4. How does one remove all air bubbles from a buret?

5. Calculate the volume of ~2.5 M NaOH solution needed to prepare 200 mL (2 significant figures) of a ~0.040 M NaOH solution. (Record your answer here and in your notebook.)

6. Calculate the mass of solid sodium hydroxide needed to prepare 200 mL (2 significant figures) of a 0.040 M solution. If one pellet of sodium hydroxide weighs 0.170 g, how many pellets should be dissolved?

7. Write a balanced equation for the neutralization of potassium hydrogen phthalate ($C_8H_5O_4K$) with sodium hydroxide. Calculate the mass of potassium hydrogen phthalate that neutralizes 15 mL of an ~0.040 M NaOH solution. (Record your answer here and in your notebook.)

8. Draw a structural representation of KHP that shows all atoms.

Date _____ Student's Signature _____

Write your answers in pen and include units.

Report Form
Analysis of a Carbonated Beverage

DATA

Beverage Sample

Volume of sample: _____ mL   Name of Beverage: _____

Initial pH of sample: _____   pH after de-gassing: _____

Solution Preparation

Concentration of stock NaOH solution _____

Volume of stock NaOH solution used: _____

Mass of KHP calculated to neutralize 15 mL of ~ 0.04 M NaOH: _____ g

Solution Standardization

	Trial 1	Trial 2
Phthalate	_____ g	_____ g
Final volume NaOH	_____ mL	_____ mL
Initial volume NaOH	_____ mL	_____ mL
Vol. NaOH consumed	_____ mL	_____ mL

Determination of Citric Acid

	Trial 1	Trial 2
Citric acid	_____ g	_____ g
Final volume NaOH	_____ mL	_____ mL
Initial volume NaOH	_____ mL	_____ mL
Vol. NaOH consumed	_____ mL	_____ mL

Determination of Beverage Sample

	Trial 1	Trial 2
Beverage	_____ mL	_____ mL
Final volume NaOH	_____ mL	_____ mL
Initial volume NaOH	_____ mL	_____ mL
Vol. NaOH consumed	_____ mL	_____ mL

ANALYSIS

Solution Standardization

                                                                   Trial 1   Trial 2

    Standardized concentration of NaOH solution: _____M _____M

    Average standardized concentration of NaOH solution: _____M

Determination of Citric Acid

                                                       Trial 1   Trial 2

    Moles of citric acid: _____moles _____moles

    Moles of NaOH: _____moles _____moles

Relationship between Moles of Citric Acid and Moles of NaOH

    Number of moles of NaOH per mole of citric acid: _____

    Number of moles of citric acid per mole of NaOH: _____

    Proposed equation for the reaction between NaOH and citric acid: _____

    Number of acidic hydrogen atoms per molecule of citric acid: _____

    Number of molecules of citric acid per acidic hydrogen atom: _____

Determination of Beverage Sample

                                                      Trial 1   Trial 2

    Moles of NaOH consumed: _____moles _____moles

    Moles of citric acid neutralized: _____moles _____moles

    Molarity of citric acid in beverage sample: _____M   _____M

    Average M of beverage: _____M

    % Concentration by mass: (assume density of 1.00 g/mL) _____%   _____%

    Average % concentration: _____%

POST-LABORATORY QUESTIONS

name _date_ _instructor_

section _partner's name_

Write your answers in pen and include units.

1. What are three of the properties that one looks for when selecting a primary standard?

2. Why is a solution of NaOH which was standardized yesterday unsuitable to be used as a standard solution today?

3. If KHP is a monoprotic acid, how would you describe citric acid? Explain your reasoning.

4. Write the chemical reaction between NaOH and citric acid.

5. Describe the meaning of each in terms of its application to acid base titrations.
  (a) standardization

  (b) indicator

  (c) endpoint

6. Describe three techniques used in this experiment.

7. Could this experiment be modified to work with a dark soda? If so, how?

Date _____ Student's Signature _____

Lab 10

PROP 2143

The Fuel in a Bic® Lighter

A Guided-Inquiry Experiment

Prepared by Vickie M. Williamson and M. Larry Peck, Texas A&M University

INTRODUCTION

When chemists prepare new compounds, they must characterize the compounds by determining such properties as melting point, boiling point, color, one or more types of spectral analyses, and/or elemental composition. In this experiment, you will investigate an unknown volatile hydrocarbon.

OBJECTIVES

You will use data obtained in lab to relate the mass, volume, temperature and pressure of a gas.

CONCEPTS

The ideal gas equation, general gas laws, pressure, partial pressure, volume, standard conditions, and the standard molar volume are several of the concepts used during this experiment.

TECHNIQUES

You will displace water in an inverted container with a sample of gas released from a disposable lighter. The method used is applicable to substances that are gases at room temperature or can be easily volatilized by heating and are not soluble in water. Your understanding of properties of gases will be used throughout this experiment. You will also make extensive use of the top-loader balance.

ACTIVITIES

You are to investigate the hydrocarbon that is the liquid you see in a disposable lighter. You are to weigh a lighter on the top-loader balance, hold the lighter under an inverted container filled with water and allow the

© 2008 Cengage Learning. ALL RIGHTS RESERVED. No part of this work covered by the copyright herein may be reproduced, transmitted, stored, or used in any form or by any means graphic, electronic, or mechanical, including but not limited to photocopying, recording, scanning, digitizing, taping, Web distribution, information networks, or information storage and retrieval systems, except as permitted under Section 107 or 108 of the 1976 United States Copyright Act, without the prior written permission of the publisher.

fuel to displace part of the water in the inverted container. Knowing the volume change in the water in the container, the temperature and partial pressure of the water, the temperature of the collected gas, and the room (or total) pressure, you will be able to calculate the volume that the collected gas would occupy when dry and at standard conditions. Then you are to study the relationship between calculated volume of gas and the change in mass of the lighter.

CAUTION

You are working with volatile hydrocarbons under pressure. Do not allow open flames to come in contact with the gaseous hydrocarbon. Wear approved eye protection while in the laboratory.

PROCEDURES

1-1. Take a disposable lighter to the analytical balance. Tare the balance with a small sheet of paper on it. Weigh the disposable lighter to the nearest milligram. Record its mass in your notebook. Hold the lighter using a small sheet of paper or fresh clean lab gloves to avoid excess handling of the lighter.

1-2. Fill a large beaker, plastic tray, or other large container with tap water. Fill a 100-mL graduate cylinder with distilled water. Place the palm of one hand over the top of the filled cylinder, invert it, and place it in the beaker or tray of water. Remove your hand. There should be no bubble (or only a very small bubble) at the top of the cylinder.

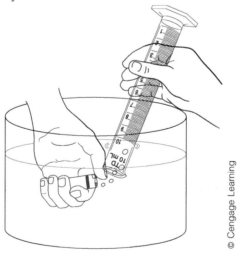

1-3. Take the disposable lighter out of the sheet of paper and hold it under the mouth of the inverted cylinder. Depress the lever on the lighter so that fuel escapes from the lighter but is captured in the inverted cylinder. Continue until the gas in the inverted cylinder is about an inch from its top. You will need to stop within the range of the volume markings on the cylinder. Remove the lighter and lay it on a dry towel or cloth.

1-4. Hold the cylinder so that it is vertical and the levels of the water inside and outside are parallel. Mark the level with your finger. Remove the cylinder. Determine the volume of the hydrocarbon that you just released.

1-5. Wipe and shake as much water as practical from the disposable lighter without causing any more fuel to escape. Take the lighter back to the top-loader balance. Re-tare a small sheet of paper and weigh the lighter.

1-6. Repeat Procedures 1-1 through 1-5 until you have four sets of data that you believe to be reliable. Do not count your first try at collecting gas. Vary the volume during each set of data; range from 20 to 70 mL of gas in your graduated cylinder. (Your first set of data will probably not be consistent with the other sets of data.)

1-7. Record the temperature of your water bath and the barometric pressure. Knowing the temperature of the water, you can go to tables such as the one below and determine the partial pressure due to water vapor. The barometric pressure at the time of the experiment will be equal to the partial pressure of the hydrocarbon plus the partial pressure of the water vapor (if the water levels inside and outside the container were the same each time the volume was noted).

Vapor pressure of water

°C	16	17	18	19	20	21	22	23	24	25	26	27	28	29	30	31
Torr	13.6	14.5	15.5	16.5	17.5	18.7	19.8	21.1	22.4	23.8	25.2	26.7	28.3	30.0	31.8	33.7

1-8. Have your instructor sign your notebook and Report Form. Using your data set, complete the Report Form.

Name (Print)   *Date (of Lab Meeting)*   *Instructor*

Course/Section

Pre-Laboratory Exercises

1. Define:
   atmospheric pressure

   vapor pressure of water

   standard temperature and pressure

   Kelvin temperature scale

2. What is the name that we give to the equation, $PV = nRT$? (Define each term.)

3. How does one obtain the total pressure of a system that is being operated at the same pressure as the current atmospheric pressure?

4. In this experiment, what measurement will limit the number of significant digits in your final calculation?

5. Draw a particle view of the gases that are present in a container in which hydrogen gas is being collected over water.

Date _____   Student's Signature _____

Name (Print)   Date (of Lab Meeting)   Instructor

Course/Section   Partner's Name (if Applicable)

Report Form

DATA

Data is collected in your notebook.

Water bath temperature: _____ °C _____ K

Barometric Pressure: _____ torr _____ atm

Vapor Pressure of Water at the Water bath temperature: _____ torr

Partial Pressure of Hydrocarbon _____ torr _____ atm

		Experimental Trials				
	Pre-trial	1	2	3	4	5
Initial Mass of Lighter						
Final Mass of Lighter						
Mass of Hydrocarbon						
Volume of Hydrocarbon						

Date _____ **Instructor's Signature** _____

ANALYSIS

	Experimental Trials					
	Pre-trial	1	2	3	4	5
Volume of Hydrocarbon at STP						
Mass of Hydrocarbon						
Mass of Hydrocarbon + Volume						
Mass of Hydrocarbon − Volume						
Mass of Hydrocarbon × Volume						
Mass of Hydrocarbon ÷ Volume						

Sample calculation for one trial:

1. Look for a pattern between the different mathematical operations (+, − ×, /). Can you identify an operation that yields a value that is nearly constant? If so, what are the units and what does the value represent?

2. Use the ideal gas equation to derive the volume of one mole of an ideal gas at STP (this is called the standard molar volume).

3. Explain how you could use a set of values like those you obtained in question #1 and the standard molar volume to find the grams/mole of the fuel. (*Hint:* Look closely at the units.)

4. Using the method you described in question #3, find the grams/mole for each of your experimental trials.

5. What is another name for grams/mole? For the values in question #4, find the average value, the standard deviation, and the relative standard deviation.

6. Give the ideal gas equation (PV = nRT). Substitute g/molar mass for the moles of gas in the ideal gas equation and rearrange the equation to isolate molar mass.

7. For each of your four experimental trials, identify the variables in the above equation for molar mass. Calculate the value obtained for molar mass for each of your four trials.

8. How does the answer to question #7 compare with the value you found in question #4? Explain your answer.

9. Given that the formula of butane is C_4H_{10}, does your answer to question #4 approximately equal the molar mass (MM) of butane? Is your value closer to that of propane?

10. What is the accuracy of your findings? (theoretical − experimental MM): _____ Find the standard deviation of your MM from the accepted MM of butane

$$\frac{|(\text{theoretical} - \text{experimental MM})|}{\text{theoretical MM}} = \underline{\hspace{3cm}} \%$$

11. Based on your answer to question #9 above, which other hydrocarbons listed below could be mixed with butane to give the average molar mass you determined?

Formulas of Small Hydrocarbons	Names
CH_4	methane
C_2H_6	ethane
C_3H_8	propane
C_4H_{10}	butane
C_5H_{12}	pentane
C_6H_{14}	hexane
C_7H_{16}	heptane
C_8H_{18}	octane

POST-LABORATORY QUESTIONS

1. Except for very small alkanes (hydrocarbons), the boiling point rises 20-30 degrees for each additional carbon atom in the molecule. Assume that the normal boiling point of the fuel in the lighter is 10°C, why was it not necessary to extend the table further (*i.e.*, why was it unlikely that your unknown contained more than eight carbon atoms per molecule)?

2. A gaseous hydrocarbon collected over water at a temperature of 21°C and a barometric pressure of 753 torr occupied a volume of 48.1 mL. The hydrocarbon in this volume weighs 0.1133 g. Calculate the molecular mass of the hydrocarbon.

3. How is the ideal gas law related to the molar mass of a gas?

Date _____ Signature _____

Supplemental Information for CHEM 1110
The Fuel in a Bic Lighter

Temperature in °C + 273.15 = Temperature in K

1 torr = 0.00132 atm

1 atm = 760 torr

$P_{total} = P_{gas} + P_{water}$

Standard Temperature and Pressure
    Temperature = 273.15 K
    Pressure = 1 atm or 760 torr

Gas Constant, R
62.4 L*torr*K^{-1}*mol^{-1}
0.08205 L*atm*K*mol^{-1}

Boyle's Law combined with Charles's Law
    $(P_1V_1) / T_1 = (P_2V_2) / T_2$

Ideal Gas Law
    $PV = nRT$

St. Deviation or $\sigma = \sqrt{\frac{1}{N} \sum_{i=1}^{N} (x - \mu)^2}$

Relative St. Deviation or RSD = 100 σ / μ